War Games

War Games

A History of War on Paper

Philipp von Hilgers

translated by Ross Benjamin

The MIT Press
Cambridge, Massachusetts
London, England

Originally published in German by Verlag Ferdinand Schöningh GmbH and Wilhelm
Fink Verlag GmbH & Co. KG under the title *Philipp von Hilgers: Kriegsspiele: Eine
Geschichte der Ausnahmezustände und Unberechenbarkeiten.* © 2008 Wilhelm Fink
Verlag GmbH

English language translation by Ross Benjamin © 2012

The translation of this book was funded by Geisteswissenschaften International—
Translation Funding for Humanities and Social Sciences from Germany, a joint
initiative of the Fritz Thyssen Foundation, the German Federal Foreign Office, and
the German Publishers & Booksellers Association.

This book was set in Stone Sans and Stone Serif by Toppan Best-set Premedia Limited.

Library of Congress Cataloging-in-Publication Data

Hilgers, Philipp von.
War games : a history of war on paper / Philipp von Hilgers ; translated by
Ross Benjamin.
    p. cm.
Includes bibliographical references and index.
ISBN 978-0-262-01697-1 (hbk. : alk. paper), 978-0-262-53832-9 (pbk.)
1. War games—History. 2. Games of strategy (Mathematics)—History. I. Title.
U310.H53 2012
355.4'809—dc23
2011026383

If on Mars there were human beings and they waged war against each other in the way chessmen do on a board, then their headquarters would use the rules of chess for prophesying.

—Ludwig Wittgenstein

Contents

Preface

Current cultural histories of the game generally exclude two spheres: the battlefield and mathematics. Yet the groundbreaking role of games in these domains could not be more serious and intensive. After the First World War, if not earlier, mathematical and military discourses in Germany not only struggled for the consolidation of their respective fields of operation, but also simultaneously discovered the game as a productive concept. From that point on, the term "war games" was no longer an odd word combination tantamount to an oxymoron. Rather, it was probably the most effective and fateful concept the twentieth century produced in order to master its crises.

It is not possible to do justice to the concept and the object of the war game without taking into consideration its long, decidedly nonlinear and not always transparent history. As a consequence, the time frame of this study, which begins in the Middle Ages and extends to the Second World War, is quite broad in scope. On the other hand, there is a clear delimitation of the area of investigation: it ranges from the medieval game boards—captured on parchment—of the German bishoprics, through the spaces of play in the baroque principalities, to the paper map exercises of the German and "Third" Reich.

A perspective that looks beyond national borders—as is often justified, if only for purposes of comparison—is here largely excluded. Instead of foregrounding relations, this study investigates quite specific constellations. The decision to highlight states of exception solely from German history seems warranted due to the fact that—from the beginning of the twentieth century at the latest—an unequaled mastery arose there with respect to both war machines and mathematics.[1]

The first two chapters begin with the medieval Battle of Numbers and extend to Leibniz's baroque symbol and machine configurations. They set forth the argument that mathematical and military semiotics could initially coincide entirely with the concept of the game and only gradually underwent a differentiation. Only in this way can it become clear that the divided mathematical and military professions of the twentieth century ultimately remain, at a subterranean level, in thrall to the game as a medium.

In particular, the design of their rule systems must be subjected to a more precise analysis. This analysis by no means excludes an examination of the permeability at the borders of their game concepts and game scenarios. Ultimately, it is also necessary to observe how the highly abstract mathematical game configurations on the one hand and the quite concrete military technical ones on the other hand merge here into the domain of general cultural technical practices.

The middle chapters are devoted to a time distinguished, above all, by Carl von Clausewitz's emphasis on the frictions of war and the "fog of war," which prompted him to reject the postulate of general calculability. In so doing, he explicitly outlined a concept of probability closely related to the game, which would first become an epistemological tool of mathematics and physics with thermodynamics. For Clausewitz, there was every reason to keep strategic and mathematical knowledge strictly separate, while traditional—and, in his eyes, outdated—military doctrine still sought to tailor the scattered operations of Napoleon's sharpshooters to rigorously geometric formations. Clausewitz's doctrine of a war of contingencies undeniably represents a milestone in the history of science because his analysis affects the concept in ways that go far beyond a philosophy of war. At the same time, however, this underscores the unsettling fact that specific epistemes emerge for the first time and exclusively in war and do not lose their force after its end. Yet one cannot do justice to Clausewitz's claim to generality when one reads him solely against his own temporal horizon, for then Clausewitz would seem to be a mere advocate of hitherto disregarded realities, which "war," in his words, is unable to capture "on paper."[2] No sooner has Clausewitz formulated this premise than it loses its validity: before long, coordination and formation systems based on signs cease to be limited to the representation of either past or possible future battles and begin to intervene decisively in steering the course of events on the battlefield. The securing of specific living conditions within arranged spaces and time frames thus appears less as a mere question of the correct use of power than as one of the correct use of the power of command. As a result, war on paper is first put into play in an unparalleled fashion. Clausewitz's military doctrine anticipates this development in a theoretical vein, but the power of command is actually implemented for the first time in the medium of the tactical war game. Not least among its consequences, the war game explodes the format of the book, that is, the very medium to which Clausewitz still entrusts his doctrines until his sudden death of cholera.

To this day, the decisive role played by war counselor George Leopold von Reiswitz in the development of this new, semiotic field of operation has not been recognized in the scholarly literature. Also pertinent in this connection is Heinrich von Kleist, who—in the course of the reforms formulated and initiated by Freiherr vom Stein—by no means only wrote plays but also engaged in war games.

After the reconstruction of the historical context—which encompasses the mathematical and military practices as much as the training in them— it will be possible in the final three chapters to focus the general inquiry on a single vanishing point. These chapters pose the question of the domain in which the operations in war and in the realm of numbers converge. That the military and mathematics have always been linked would not be a new claim.[3] However, the lines of connection have hitherto been drawn primarily in the domain of technical achievements. Mathematicians seek to advance such achievements and strategists attempt to make use of them. But if one takes the game as the linking element, it is possible to delineate a space that has not always already been determined by a teleological factor. Rather, the game turns out to be a site from which military and mathematical practices first arise, even before concrete applications are able to justify them. Thus, it is necessary to demonstrate that the mathematical discourse of the 1920s was polarized into formalist and intuitionist positions only on the surface, via the substantiation or rejection of a mathematical metalanguage. Below the surface, however, with the concept of the game, a metalinguistic object had long since prepared a common ground for the controversies.

The war games of the Reichswehr, on the other hand, show what parameters are required for regimes to erect their concrete power structures on the basis of these paper operations. A special function is thereby assigned to war games: construed as media, they provide information about a historiography in the mode of the General Staff. This historiography has itself become part of military technique. It no longer derives claims to power from the past, but instead—in close connection with map exercises—secures access to immediately pending time periods. Thus it will be necessary to take into account a double contingency: a contingency framework is embedded in the war game, and the incalculable breaches of this framework—which occur in the course of the games—have the most decisive consequences for real military command structures.

The study of war games calls for a critical engagement with game theories and media theories, which set the fictional and the simulation in opposition to reality. The sociologist Jean Baudrillard, for one, long ago announced the dawning of the age of simulacra. In his analysis, simulacra can no longer even be conceived as the appearance of reality, but instead establish themselves through self-referentiality. In opposition to this sociology stands a history of war games—and thus of simulations—that have not been subsumed in absolute virtuality. Instead, they have foundered on stumbling blocks of all sorts. But it is precisely through such failures that war games unleashed a peculiar form of productivity.

The game configurations under investigation should be conceived as techniques through which subjects first constituted themselves. In particular, mathematicians at the beginning of the twentieth century could still believe that they belonged to a discipline that was suspected at best of "playing games."[4] Yet this actually enabled them, rather inconspicuously, to design the fields of operation for the Second World War. With a focus on John von Neumann as the founder of game theory, that is the topic of the concluding chapter.

Acknowledgments

"Kriegspiel" is among the German words that have found their way into English (even though it lost an "s" en route from one language to the other). This book is an attempt to carry more of the term's context across the gap between the languages. For their help with this transfer, I would like to thank the initiative *Geisteswissenschaften International*, Friedrich Kittler, Raimar Zons, Peter Weibel, and Stefan Lätzer on the one end and the German Book Office, David Mindell, Geoffrey Winthrop-Young, Ross Benjamin, two reviewers, and Marguerite B. Avery on the other.

1 The Battle of Numbers in the Middle Ages

Formations of the Battle of Numbers

According to Adam Ries, it is necessary to distinguish between "calculation on the lines and with the quill": numbers can be positioned as counters on the lines of an abacus, the antique calculating board, or they can flow in the form of Hindu-Arabic digits from the quill.[1] But when Ries extolled the virtues of writable digits in the early modern period, he did so in a medium that did not stand in a neutral relation to the represented numerical concepts. Gutenberg's book printing preserved and reproduced writing operations better than it did anything else. When the Occidental and Oriental forms of calculation first encountered each other in Italy and Spain in the Middle Ages, it was not merely different modes of representing calculative operations that came to the fore. Rather, it turned out that the numerical conceptions differed at all levels of their material incarnation. The most dramatic difference emerged in the comparison of their place-value systems: Whereas on the abacus—the *tabula abachi*—the place that does not count is simply not incarnated by a stone, the Hindu-Arabic numeral system indiscriminately indicates a value and the lack of the same through signs. The news of zero is therefore placed by some authors, with a certain degree of justification, at the beginning of the history of the modern period.[2]

The history of the Battle of Numbers, however, first created a platform on which various mathematicians were able to enter into competition.[3] What began in substance in the eleventh century received its name in the twelfth: *Rhythmos* and *machia* were combined by clerics into *Rithmomachia*[4]—a coinage in which the first lexical component not only means *arithmos*, or "number," but is also read as a musical quality. Yet the Roman Boethius had uncoupled mathematics from music in the sixth century when he established the very numerical proportions from which the Battle of Numbers now derived its configurations. Cassiodorus took still further the separation of numerical conceptions from all "material supplementation"[5]—the division of the quadrivium, which was based on different applications, seemed invalid to him, and he promptly summarized it as *mathematica*. The Battle of Numbers, however, again sets in motion an operational approach to arithmetic. By bringing the confrontation of even and odd numbers onto the game board, the Battle of Numbers aligns with the basic concept of Pythagorean mathematics.

Initially, the term Battle of Numbers was not associated with the attribute of play. Only relatively late is there mention of *ludus*[6] in connection with the *conflictus numerorum*.[7] In light of the conflicts at the level of numerical practices, which were fought out with the Battle of Numbers, one cannot be certain that its limits are those of a game. The contrast with and distance from the pure game becomes conspicuous, at the latest, through its reception in the baroque period. In its collecting mania, that age takes up the Battle of Numbers as nothing more than a scarcely understood game with mute signs.[8]

Yet the *Rithmomachia* is probably the first instrument that is not only described in writings, but also emerges from writing itself (figures 1.1 and 1.2). One searches in vain for diagrammatic designs of this complexity in previous epochs. The Battle of Numbers disseminated its forms of inscription with a comprehensiveness that erases the difference between writing and calculation, at a moment when the written calculation of Arabic mathematicians found its way into Western Europe.

One of the most prominent figures among the scholars of the twelfth century, Hermann the Lame, assigns the Battle of Numbers to the arsenal of medieval instruments—including the astrolabe, the abacus, and the monochord—and stresses its instrumental character.[9] It thereby serves primarily as a means of practice in figuratively understood numbers. The goal is to arrange one's own pieces on the opponent's side of the game board in accordance with the proportion doctrine of arithmetical, geometric, or musical harmonies. The calculation and game principles coincide with the mathematical founding acts of the Pythagoreans and to this day pose riddles to archeologists and philologists in their attempts at reconstruction.[10] Nonetheless, the Battle of Numbers is distinguished from the astrolabe and monochord by the fact that it does not refer to external realities such as stars or

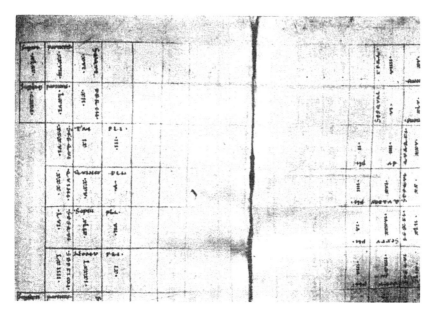

Figure 1.1
The oldest known depiction of the game board for the Battle of Numbers, prepared for the cathedral school in Hildesheim around 1100, accompanying instructions by Odo von Tournai.
Source: Bistumsarchiv Trier, BATr Abt. 95, no. 6, fol 79r. Reprinted with permission.

sound images. And as for the abacus, it is employed for a whole variety of practices: it serves merchants as much as geometricians.[11] The Battle of Numbers, on the other hand, turns the translational achievement of the abacus on its head. As opposed to the abacus, which has as its only object calculation itself, the Battle of Numbers incorporates more and more symbolic and objective contexts in the course of its development: musical intervals, battle formations, and thus whole world orders are enacted in the Battle of Numbers, without particular figurative and iconic efforts being undertaken in the process. In the manuscripts of the Battle of Numbers, which were produced for over six centuries at least, the game pieces are rarely described through colors and geometric shapes. The Battle of Numbers is surprisingly symbol-laden for an epoch in which the imaginary reigns above all. Unlike in the case of chess, for example, to this day no game board has been found for the Battle of Numbers. This proves *ex negativo* that the Battle of Numbers was bound only to the possibilities of the medium of parchment.

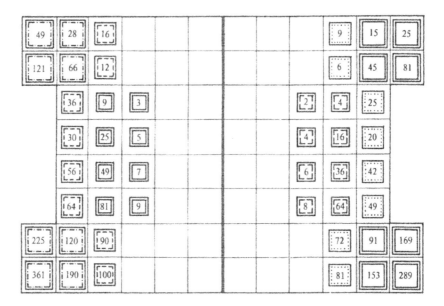

Figure 1.2
Game board for the Battle of Numbers reconstructed by Arno Borst according to the
Liège table with even and odd game pieces derived according to various classes of
proportions.
Source: Borst 1990, 278. © 1990 WBG. Reprinted with permission.

The Implantation of Mathematics

Arno Borst has reconstructed the discursive milieu of South German cathe-
drals around the year 1000, within which the Battle of Numbers arises. The
catalyst was the so-called Worms school quarrel. The two cathedral schools
of Würzburg and Worms struggled for the favor of pupils and ultimately
for that of the Salian Emperor Conrad II himself. By itself, the quarrel
would not necessarily have led to a retreat from the principle of orally
competitive rhetoric. But apparently the Emperor's chancellor and cousin
explicitly decreed that it should be fought out in writing, and a monk
named Asilo came up with the idea of composing a Battle of Numbers.[12]
The cause of the quarrel itself—the efficient calculation of sums of arbitrarily long series and setting up of ratios—favors the writing surface and
evokes forms of inscription. Early commentators already characterize the
Battle of Numbers as a *novellae plantantiones*.[13] It makes possible a tentative

writing,[14] which gains traction through an arrangement that repeatedly evokes new orders: Begun in the form of a circular letter and continued in composite manuscripts, the scattered writings on the Battle of Numbers nonetheless escape all luxury volumes and canonical writings.[15] The "disposable literature"[16] in which the number conflict is fought out does not flow into the dogmatic stock of knowledge. One exception, however, seems significant: in a single case, comments on the Battle of Numbers are taken up in a luxury manuscript alongside venerable texts on the *regula* and *ordo* of the monastic discipline. Whether this exception rests solely on a mistake—provoked by the frequent use of the signifier *regula*[17]—or whether a space for play is in fact being granted in the enumeration of monastic rules is an open question.

What unites and divides the three introductory and four additional liberal arts of the Middle Ages is their use of letter-based or numerical sign systems. Only the focus on the use of writing characterizes all the subjects of the *artes liberales*. If a secure logic of counting is first inherent in Roman numerals, it is still possible for Greek letter-numbers to make what is counted nameable through the alphabet. The simplicity of that which can be straightforwardly announced and said could always be elevated to the last explanatory resort[18] alongside that which can be geometrically shown in Pythagorean mathematics—especially as mathematics and music theory are linked down to their technical terminology.[19] However, the Greek sources became more and more linguistically inaccessible to the Western empires of the Middle Ages.[20] Increasingly, therefore, it was possible to perform operations with Greek signs only as such. In Greek letters, Carolingian monks discover the link that translates orders of writing into numerical orders: in the cryptograms of the papal couriers, names can be encoded through numbers, and sums that yield names written in Greek open up—beyond all calendrical calculations—a glimpse of looming apocalyptic events.[21] Tangibly practiced arithmetic nonetheless differs fundamentally from its inscription up to the first millennium: whereas monochords, sand tables, wooden abaci, their *psephoi* and *apices*, and even finger positions took on the most diverse spatial and temporal configurations, the act of setting them down in writing leads to orders of inscription that are bound to the direction of reading and writing and are ultimately immovable.[22] Until the appearance of the Battle of Numbers in the eleventh century, there are—as far as can be ascertained—no instances of movable and discrete elements that exhibit numerals and do not arrest their arrangement. Rather, established numerical designations refer on their writing substrata directly to movable elements—for example, the signless counters of the abacus or the strings of the monochord—in a continuous, sequential fashion. That writing in the mode of continuity does not constitute a triviality first becomes clear with the onset of Arabic algorithmic notation: The backward movements of reading, the space-seeking directions of writing, the cross-outs—undertaken by reading and writing operations in rapid alternation on discrete signs—are all basic in themselves. But no one had previously been compelled to take them up. Conversely, a prominent passage by Herodotus demonstrates that the use of the abacus follows the movement of writing: "In writing letters and in calculating with stones the Greeks move the hand from left to right, the Egyptians from right to left."[23] The Battle of Numbers will first systematically open up further dimensions of the field for semiotic operations through horizontal, vertical, and diagonal ways of moving the game pieces and calculating stones. It will stack signs into pyramids and raise them from the surface into the spatial realm. In short, due to the loosening of the grip that prescribes the direction of writing, multidimensional spaces open up, in which sign systems are subjected to an elementalization. Doctrines of the abacus limit the movement of the counters to specific axes, lest the logic of the place-value system be thwarted. In the Battle of Numbers, on the other hand, there are three interconnected levels that can emerge as numerical representations: What counts equally and simultaneously are the fields of the chess-like game board, the number of the game pieces and the numerals on the game pieces. The Battle of Numbers ceases to function as an instrument for calculating numerical relations. It is not as much about numerics as it is about numerology—the maximization of numerical relations and referents, not the calculation of quantities. The Battle of Numbers skillfully limits the calculation of numerical relations: only pieces with low numbers can be combined into a large number of products and sums that correspond to the pieces with the highest numbers and can thereby win. Conversely, for pieces with the highest numbers, only division can be used to eliminate pieces with lesser numbers from the field through one of their divisors.

   The high density of arithmetical relations that the Battle of Numbers produces must be managed with mental calculations. Increasingly, tables of ratios are available to struggling players, and the Battle of Numbers degenerates—to its inventors' chagrin—into a war of tables. The numerals of diverse cultures and epochs find a playing field in the Battle of Numbers. A battle for supremacy of the various numerical concepts is literally fought out here: Roman, Arabic, Greek.[24]

One disadvantage of Roman numerals clearly exposed by the Battle of Numbers is that with higher numbers, they tend to require a great deal of writing surface, which is just as hard to apply to game pieces that are all the same size. But Greek letter-numbers and gobar digits—to an equal extent—might have first demonstrated that scalarity could also be applied to numerals and that—in the case of gobar digits—the directions of writing or reading could shift. The Battle of Numbers stands at the intersection of a decoding of the sunken numerals of the Greek and Roman epochs and of the future ones of the Orient.[25]

"Caracteres"—a new term that emerges from this juncture—implies the dissolution of the strict separation between written numerals on the one hand and the operationality—in itself devoid of characters—of the instrumentariums on the other hand. From that point on, numerals achieve autonomy in the course of abiding traditions of writing. Meanwhile, their instrumental implementations in the form of the abacus and other calculative apparatuses have long since disappeared. Their reconstruction becomes a speculative question. And so scholars of the Middle Ages train themselves for the first time in mathematical descriptions, for the understanding of which the materiality of parchment suffices.[26] Even before the turn of the millennium, Gerbert of Aurillac did not simply presuppose the abacus in his *Regulae de numerorum rationibus*. Rather, he completely redesigned it, in order to practice the numerical relations that appear in the sentences of his source.[27] One reason that the calculating stones can no longer be presupposed is that they become a hybrid construct on which the stamp of writing is imprinted for the first time; in order to provide them with gobar digits, Gerbert ordered that they be fashioned out of horn.[28] Caracteres thus designate very precise numerals, which for the first time appear on the side of mobile elements like game pieces and calculating stones. The crossings of the place-value systems that thereby occur might have initially produced incalculabilities above all. But beyond that, a combinatorial matrix with movable letters emerges, on which—not least of all—the Gutenberg Galaxy will be based.

Scholars are divided as to whether the Battle of Numbers does not already arise in Walther von Speyer's *Libellus Scolasticus* of 984.[29] A personified geometry begins here as "a playful battle"[30] with the above-mentioned caracteres. However, columns of the abacus numbers one and ten dominate the event, and not—as with the monk Asilo half a century later—Boethius's classes of proportions. Nonetheless, Walther condenses—in the form of dactylic verse—numerical proportions, calculative operations on the lines, numerical figures, and musical interval formations into the program of *mathesis*. In the development of the Battle of Numbers, everything that still sounds metaphorical here will take on a calculable and playable form on the same written basis.

Semiotic Turn

What is the status now of the fragility of things, the persistence of the grapheme and material and semiotic transferences? Regarding the partition of the pieces on the game board, the first writings on the Battle of Numbers reveal nothing; nor do they provide any game diagrams. Nonetheless, the first extant tabular arrangements of the pieces show at a glance a highly differentiated grouping. Their schema follows exactly Greek military formations.[31] The pieces are permitted to move in different increments. With each move they travel one, two, or three fields.[32] It is as if heavily armed hoplites, more mobile foot soldiers, and riders were waging their attack on the wings of the game board. To think strategies and numerical figurations together is a Greek achievement.[33]

With the Battle of Numbers—despite or precisely because of its abstraction—religious scholars brought in a military reality. Roman war chronicles already spoke of their armies as of signs: Thus, phrases such as "signa provere" and "signa constituere"[34] stand for the advance and halt of whole troop units, which are themselves no longer addressed. The Roman military counted among the "signa" not only flags, but also acoustic signals. Specific chords of individual horns had only a single addressee—the sergeant and standard bearer, the *signifer*. He translated the acoustic signal sequences into optical ones.

The eleventh century, in which the Battle of Numbers arose, appears to have drawn from such sources of the use of signs. According to Carl Erdmann's investigation into the emergence of the "idea of the crusade," it is reflected less in a Christian iconology than in semiotic practices that are typical of medieval battlefields. Thus, a theosophy became possible that no longer ethically condemned or justified wars, but itself created reasons for war. Erdmann's attention is therefore directed primarily at the holy flags that arose at the turn of the millennium.[35] With the beginnings of the Christian sense of mission, the *ordinatio*—the power of consecration—established the hierarchy of the Church, separating bishops from priests, priests from the laity, and sacred objects from profane things. But only in the eleventh and twelfth century was a boundary crossed in the semiotic orders: the consecration of flags and swords assigned insignia of a military order to the churchly order. Strictly speaking, flags had hitherto exhibited a trinity that profoundly opposed the Christian one. Flags were not only incriminated as lance weapons and—still more devastatingly—through images of idols. On top of that, they counted among the signa—the standards. As such, they made the battle and combat legible; they regulated beginning, middle, and end. They were no longer separable from the war that they waged. Chiastically, the Church designed its own flags, provided them to the armies, and—conversely—led crusades under the flags of kings. The battle was no longer waged merely with signs but over signs. Depictions and miniatures of the crusades differentiate the often completely similar Franks and Saracens on foreign and unknown ground solely by the fact that the former displayed insignia and the latter did not. The victorious end of the battle was sealed with the reconquest of holy flags by the king who captured them.

From that point on, signs gained an autonomy of previously unknown magnitude. Probably unsurpassed in this regard was the *carroccio*, a wagon bearing the standards of those Lombard cities that preserved their independence in 1176 in the victory over Barbarossa. Before each battle, the *carroccio* was fetched from the cathedral by a city contingent made up not only of soldiers, brought to the marketplace, equipped with all sorts of insignia and finally taken to the battlefield. During the battle itself, a group of guards protected the wagon, while on its platform trumpeters sounded tactical signals; notaries wrote orders, recorded losses, and prepared commendations, punishments, and compensations; and priests cared for the wounded and administered sacraments to the dying. "Thus, the classic *carroccio* served several purposes at once for the northern and central Italian city communities: as a sort of mobile generals' hill, command center, optical point of reference," as a dressing station and refuge for weary soldiers.[36] Above all, however, the wagon bearing the standard ensures a self-contained war, because to capture it means to take possession of the *signum civile*—without extending the battle to the city itself.

In the Battle of Numbers, one game piece—the pyramid—is now elevated above all the others. It embodies several square numbers at once. The taking of all the other game pieces is executed through expressions of arithmetic. But the taking of this piece is articulated only through a military terminology.[37] If the pyramid—which is vulnerable in comparison to other game pieces—is taken, then all the other pieces that count among the square numbers of the pyramid are rendered invalid.[38] No other piece contains such purely referential dependencies. The rules of the emerging chivalric orders will provide the same semiotic logic for the battle: if the standard bearer falls, then the troops assigned to him admit defeat as well.[39] Thus, the Battle of Numbers overlaps with the rules of the chivalric orders and has, so the theory goes, created a codex for their peculiar position midway between military and clerical status.[40]

2   Power Games in the Baroque Period

Spaces of Play

Of all centuries, it was the seventeenth—which engendered reason and assembled mathematics into a discipline from the obscure semiotic practices of secret societies and the semiotic regimes of ideal states—that found in games an epistemic reservoir. Gottfried Wilhelm Leibniz led the way in discovering in games a playing field of knowledge. The space that games occupy in his work does not serve allusions and allegories. Rather, it is characterized by its own genuine technicity and materiality. It is precisely games that are assigned the task of revealing the universality of cultural techniques such as measuring and drawing, calculating and combining— indeed, primarily in the limited space of the book. Sign systems emerge that not only describe the elements at play, but also implement them operationally and thereby carry them further. Books thus reveal playing fields of action and signs that can be taken up by other books without having to draw, in exegeses and commentaries, from a source of authorized discourse. But the interoperability that transplants the game into texts with signs and graphic elements does not form a closed system of the text. Rather, it establishes within texts platforms from which objects and artifacts first arise.

In the games of the seventeenth century, representational forms suffer a breach. In their place, semiotic operations are promoted to the prosperous switch point of knowledge. Games are themselves released from purposelessness. They can change at any point into a teleological model entrusted even with foregrounding underpinnings of the state: Fortifications and theater buildings, firearms and fireworks, or mathematics and games are skills that find representation in the very same books.[1]

Still more than games, it is necessary in what follows to keep game boards in mind. It is to these that Leibniz repeatedly has recourse for the development of his *ars characteristica*. The core of his *ars*, however, is the production of objective and worldly contexts, which unfolds on and through paper. Their test is to be assigned to a calculus. But Leibniz's program should not be read simply as a progression of increasingly abstract relations between signs that turns away from existing languages and toward mathematical notations. Rather, the question is what was lost or had to be lost before scholars—since the nineteenth and twentieth century at the latest—saw in Leibniz's writings a reductionism at work that they took up and carried further, only to come ultimately to nothing but circular arguments in this program.[2]

The *ars characteristica* may indeed be based on two arts, which Leibniz conceived as *ars inveniendi* and *ars iudicandi*. In substance, however, he intertwined two lines of development that had found their modus operandi in operations with letters. The algebra of François Viète and subsequently that of René Descartes managed to reduce geometrical figures to calculations with letters. And secondly, Leibniz himself—in his dissertation on the *ars combinatoria*—had pursued the systematic decomposition of words, which had likewise revealed a basic operational element in letters. Conversely, words and even neologisms can emerge synthetically from permutations and variations of letters—just as geometric and hitherto unseen entities can emerge from algebraic calculations. If the latter—the production of new objective contexts—was the task of the *ars inveniendi*, the *ars iudicandi* had to subject to a calculus not only the consistency of the decomposition of existing words and geometric images, but also the process of their new creations. In the final analysis, every establishment of truth thus amounted to the proof of a flawless calculation.[3]

The Renaissance had already produced diagrammatic constructions that went beyond mimetic relations between art and nature and displayed mathematical functions. Leon Battista Alberti collected them in a book that, significantly, declared the game the object of mathematics.[4] Here the "clever bombardier" learns how he can measure the angular distances of remote objects with the help of a planisphere and calculate the proper alignment of his cannon muzzle with a pendulum. The mathematical instruments also served Alberti in the more pleasurable task of mapping Rome.[5]

Samuel Edgerton goes so far as to assume that, in the modern period, it was due to perspectivist techniques of representation that it became possible to develop constructions of machines and lever mechanisms on paper alone.[6] This thesis is contradicted by the fact that no new machines were actually designed in this time period and the forces inherent to the machines could not be represented with the method of central perspective.[7] Moreover, adepts employed discursive strategies to draw their knowledge and their power from the correct application of books. It was necessary, however, to retain the key to their operation at all costs—for example, through display of the geometric solution and concealment of the algebraic process of calculation.[8]

More cautiously formulated, it can be said that the apparatuses that the modern period invented in its books were optical apparatuses that disseminated and differentiated methods of representation.[9] Only when it came to perspectivist constructions did books achieve a previously unknown self-sufficiency, which culminated in the case of games. The explanation of drawing techniques already availed itself of auxiliary visual constructions in its argumentation. It also recommended necessary construction aids such as proportional dividers and triangulators for reproduction and ultimately demonstrated the targeted effect in pictures.[10]

In particular, the figures in books on theater buildings venture to represent the perspectivist methods of illusory architectures with those very methods, in order to demonstrate how stage spaces should be constructed and how, through scenery painted in a perspectivist fashion, they can be endowed with the illusion of a nonexistent spatial depth.[11] Diagrammatic hybridizations are demanded here that, to a certain extent, disrupt the imaginative effect of pictures by means of letters and render them identifiable as a construction. Algebra supposedly emerged from just such abbreviations, which designated specific geometric elements of the figures and then became an object of mathematics themselves, thereby separating general procedures from concrete problems.[12] The translation of antique texts on geometry and arithmetic was accompanied by their fundamentally new visualization. Mathematical texts of Greek origins reached the Western world without figures and diagrams.[13] Algebra did not merely pave the way to converting pictorial relations into letter relations. On the contrary, it also—by circumventing the descriptive and symbol-free prepositions and argumentations of the Greeks—enabled new pictorial and representational procedures to emerge from pure letter relations.

Leibniz ultimately expected algebra to accomplish the design of machines straight from paper, without relying on figurative and perspectivist representations:

I can represent with characters and without figures or models extremely intricate machines, as if I had drawn them and designed them in a model; or even better than that, for with this symbolic representation I can calculate, as it were, shift and change the machine on paper and seek the correct positions through analyses, whereas I would otherwise need countless models to do the same, and on a trial basis.[14]

But Leibniz by no means revokes the relation to pictorial space. On the contrary: for Leibniz, the condition of possibility of a "blind thinking" commences with algebra, which is relieved of presenting objective relations. It borders on an elimination of "intellectual work," because "arguments" obtain their conclusiveness "due to material data." Instead, thinking consists of seeing a thread "which is perceivable with the senses and which, as it were, mechanically leads the mind, so that even the dumbest can follow it," and thus "the truth can be reproduced and as if with a machine printed and captured on a piece of paper."[15]

Instead of merely pursuing deductions that can be drawn from Leibniz's semiotic abstractions, this argument opens up possibilities for concretizing semiotic realizations more sharply. For it is not only on a stage transferred into the mental realm that logical constructs collide. They already do so on the material substratum, which can be captured through recording and inscription techniques that simultaneously belong to it.

Leibniz's Graphemic Strategies

The Middle Ages knew seven liberal arts, which covered all the skills of speaking, writing, calculating, showing, and drawing. The index that merely begins to take into account Gottfried Wilhelm Leibniz's still extant 75,000 writings and 15,000 letters[16] could be considered a register of the seventeenth century, insofar as the epoch was embodied to the highest degree in Leibniz: who could count all of the more than 150 arts before he finally arrives after over six columns at the *ars vivendi*?[17] In Leibniz's register of the arts, the *ars inveniendi* occupies a special place because it is the root of all arts. Already in his lifetime, Leibniz's tentative development of the *ars inveniendi* led to a vast abundance of papers and collection of artifacts as well as a large number of scholarly institutions and correspondence networks. All of these products taken together raise the question of what else the completion of his art of invention—repeatedly called for but never attained—could have actually yielded scientifically.

Leibniz developed his *ars characteristica* not only through arrangements of letters. Increasingly, he also brought in two-dimensional graphic-geometric frameworks, such as topological tree structures, various networks, or quadratic area divisions. Helmut Schnelle has scrupulously enumerated all the graphemic operators—at a time when cybernetics was poised to traverse virtually all the sciences.[18] He noted not without surprise that the graphemes were not readily extracted from the extant sources.[19] In the Leibniz literature—in which the liberation of the metaphysician of reason from an epoch of occult semiotic practices has top priority—there are only scattered indications that Leibniz is indebted to games for some of his fundamental mathematical principles and graphic arrangements. In his commentary on Johannes de Sacrobosco in his first publication, the *Dissertatio de arte combinatoria*,[20] he consults—alongside Clavius's combinatorial deliberations—above all Georg Philipp Harsdörffer and Daniel Schwenter's *Deliciae Mathematicae*.[21] And at the end of his career, he still expects from the mathematical analysis of all known games that bear some relation to numbers the realization of his *ars characteristica*, a task that he emphatically advises the mathematician Pierre Rémond de Montmort to undertake.[22]

The nineteen-year-old Leibniz first entered the scholarly mathematics of his time with his *ars combinatoria*. The writing coincides with a break that characterizes the teaching of mathematics in general in the middle of the seventeenth century. Thus, Harsdörffer's *Deliciae Mathematicae* opens up for mathematical practices a field beyond that of the drill of primary schools and the business of merchants' schools.[23] But even when Harsdörffer takes up the work of the linguist and mathematician Daniel Schwenter, his poetological elaborations are closer to the *inventio* as part of rhetorical doctrine than to the emerging praxis of engineering. Techniques of compilation likewise still entirely serve writing, and Leibniz will be the first to derive from that the combinatorics that helped bring the *mathesis universalis* to a central epistemic position.

Even if Harsdörffer's *Deliciae Mathematicae* cleaves to paper and poetics, it is for this very reason that it summons new forms of mechanization: the bookbinder is instructed to cut up a piece of paper with the figure of the fivefold *Denkring* (thought-ring) of the German language into the same number of rings, to mount it on firmer paper, and finally to affix it concentrically and rotatably (figure 2.1).

The fact that wheelworks—"ex papyro"—can henceforth be components of books does not escape Leibniz in his *ars combinatoria*.[24] And, as will be shown, he will know how to use Harsdörffer's mathematical recreations to wage a public campaign. The course has already been set by Harsdörffer. He does not seem to have derived the construction of the German *Denkring* from the diagrams of the *ars magna* by the Catalan monk Raymundus Lullus. Rather, he adheres to a model by the Huguenot military writer Sieur du Praissac de Braissac: "Briefve méthode pour resoudre facilement toute question militaire proposée."[25] Du Praissac's idea of achieving strategic measures with the help of applications might itself have been inspired by Moritz von Nassau, whom he accompanied on his campaigns as a reporter. Moritz and Ludwig Wilhelm von Nassau are demonstrably among the first to draw their battle formations from Greek—and, of course, nonpictorial—sources and test them in war games.[26] In particular, the invention of linear tactics can be traced back to Wilhelm Ludwig von Nassau, who proposes in a letter to his cousin Moritz the principle of rotating musketeers, who—positioned in five rows of nine—always advance one row during the loading of their firearms, and finally, after the shot has been fired, reposition themselves in the last row. Linear tactics provided a higher continuity of salvos and simultaneously granted the musketeers better protection in the moment of reloading. Of all this, Wilhelm Ludwig's notepaper contains nothing more than the rule system of a cyclic alternation of letters that it was necessary to inscribe on the soldiers as discipline (figure 2.2).[27]

Du Praissac's application stands for the attempt—analogous to Wilhelm Ludwig von Nassau's tactical arrangement—to affix strategy to a rotating mechanics. Here, an inventory of questions of warfare is systematically gone through. Ultimately, Harsdörffer's *Denkring*, which undertakes to "show the whole German language on one piece of paper,"[28] merely transfers—through its recourse to du Praissac's template—syntagmata of the battlefields into the realm of the German language. No less committed to

Figure 2.1
Philipp Harsdörffer's fivefold *Denkring* (thought-ring) of the German language with instructions on its installation within the book for bookbinders.
Source: © 1990 Keip. Reprinted with permission.

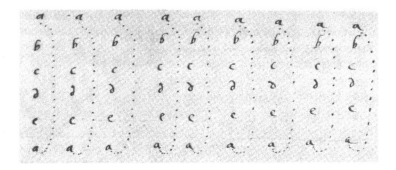

Figure 2.2
Design of "linear tactics" by Wilhelm Ludwig von Nassau, 1594.
Source: The Hague, Koninklijk Huisarchief, MS. A22-1XE-79. Reprinted with
permission.

this poetological program, Justus Georg Schottelius, the Wolfenbüttel linguist who was his friend and colleague, describes the decomposition and
construction of the German language as a "terrible language war"—a consequence of the Thirty Years' War, as it were.[29]

The *ars combinatoria* had been intended to earn Leibniz a professorship
at the University of Altdorf, where Daniel Schwenter and Philipp Harsdörffer had worked. But Leibniz broke out of the academic circle and
famously chose to travel to Paris in the diplomatic service of Johann
Philipp von Schönborn, the elector of Mainz, with a plan of attack against
Egypt. The objective was to redirect the power interests of Louis XIV from
Central Europe to Egypt.[30] There, it was not the design of a calculating
machine that he brought with him as an admission ticket to the Académie
des Sciences, but a plan that would provide proof of his juristic and diplomatic suitability—with which he famously failed. Leibniz had, however,
previously tested his diplomatic skills under the aegis of his sponsor, minister to the elector of Mainz, Baron Johann Christian von Boineburg. His
effort is worthy of closer scrutiny.

When the King of Poland, John II Casimir, abdicated the throne in 1668,
the tsar's possibilities of influence in Central Europe threatened to over-
power the Electoral Palatinate in the choice of the claimant to the throne.
A rival candidate was to be placed on the vacant throne. Leibniz attempted
to demonstrate through a syllogistic process that no one but the palsgrave
Philipp Wilhelm von Neuberg would be eligible. The British economist

John Maynard Keynes saw in Leibniz's writing the beginnings of a new logic comprising the doctrine of probability.[31] German logician Heinrich Scholz disputed Keynes's argument, asserting that Leibniz merely applied the traditional syllogistics to a new field.[32] In fact, however, Leibniz seems to have taken up du Praissac's method, which he knew from Harsdörffer. According to this method, if a question made up of truisms is answered in the affirmative, then it constitutes the point of departure for a series of subsequent questions produced through the corresponding turn of the rings of du Praissac's circular schemata: "If war has now been decided"— that is, if the question of "whether one should wage war" has been answered in the affirmative—"then one must hold together the question of the first and fourth rows to consider whether one shall remain, whether one shall yield, whether one shall battle," etc.[33] It is precisely according to this con- catenating schema—which du Praissac did not regard as limited to military application[34]—that Leibniz's *catena definitionum* proceeds,[35] in order to come to the conclusion that the Palsgrave von Neuburg is the only legiti- mate claimant to the Polish throne.

Christoph Weickmann's Power Game

In 1616—three years before the foundation of a new science appeared to Descartes in a dream, a method that he would spell out in his "Rules for the Direction of the Mind"—a sentence appears in the great chess book by the future Duke August of Braunschweig-Lüneburg, stating that physics "lends matter to numbers, masses and divisions: though in this game matter can be excluded by the intellect, along with a good memory, when it is firmly imagined in the same ."[36] It would scarcely have been possible to prefigure the diverging course of the *res cogitans* and *res extensa* more radically than Duke August did: from now on, bodies may "drive, ride, or walk," while the intellect pursues "by rote" all possible "courses and moves" of a chess game—which is, however, admittedly "rather hard to set to work."[37]

After the Thirty Years' War, in 1664, the Ulm patrician and merchant Christoph Weickmann had a dream himself: after a day of extensive games of chess, a game appeared to him in his sleep, liberated of all external *objectis*, in an entirely "new form" and "figure."[38] Instead of the quadratic fields of the chess board, a network made up of nothing but straight and intersecting lines formed the basis of the game. Weickmann set the game down on paper and published it as a "Newly Invented Great King's Game." His book not only relied on August's chess book in its title, but was also dedicated to him. The writing was divided into two books. The first one reveals the external nature of the game and its rules. The game offers the second book a pretext to make sixty "observations" with baroque prolixity, from which—after various historical examples and numerous authorities— regimental and military rules are ultimately deduced. The first book, which constitutes less than a sixth of the total writing, provides information about the production of the game, its figures, the ways the pieces move and take one another, and the game's objective. This last aspect amounts, as in chess, to placing the king in checkmate. The production of the game boards is no longer left to a bookbinder, as with Harsdörffer's *Deliciae Mathematicae*, but is now assigned to the reader. Four different game boards are to be transferred from copperplates onto firm paper and to be mounted on wood, though the scale must sometimes be doubled or tripled.[39]

The four game boards make possible a game with two, three, four, six, and eight players. Instead of the sixteen figures of chess, from which Weickmann explicitly derived his game, the players in his version each initially have thirty figures, to which fourteen different ways of moving are assigned. Circles mark the figures' positions, and lines the directions of the moves. Whereas in chess a field that is not on the edge always borders eight others, Weickmann does not connect all adjacent fields. Rather, his network consists of elements that are connected alternately in fours and eights. He divides the lines of connection into two different classes of diagonal and orthogonal lines and instructs the reader to color them differently. With the topological configuration of the board, which reproduces graphically and marks with signs and colors not only the figures' positions but also the moves themselves, ways of moving become diagrammatically addressable. If in chess possible moves are provided only by the figures, in Weickmann's game the board provides various possibilities for moves and forces certain figures on predetermined courses (figure 2.3).

If Duke August in his chess book, for the amusement of the reader, still mentioned chess figures that bear the insignia of court dignitaries, Weickmann in his tableau explicitly equates faithful pictorial depictions of officials, game figures in the floral forms of baroque woodturning, and astronomical signs, which are found in the illustrations of the game boards for the arrangement of the figures (figure 2.4).

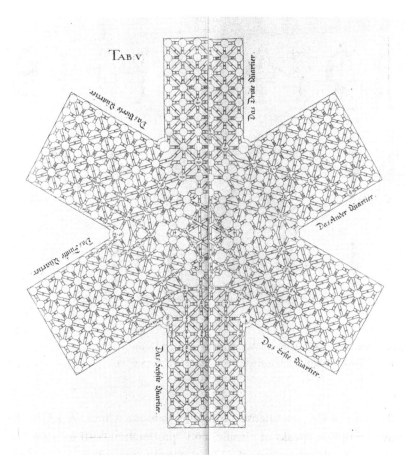

Figure 2.3
Game board of Weickmann's "King's Game," Ulm 1664.
Source: Herzog August Bibliothek Wolfenbüttel: A: 5.6 Pol. 2°. Reprinted with permission.

Weickmann's purpose with his game is not so much entertainment as the attempt to derive from it a "state and war council," whereby "the most necessary political and military axiomata, rules and ways of playing . . . without great effort and the reading of many books, are shown and presented as if in a compendio."[40] It may well be a consequence of the Thirty Years' War that the figure of the king is surrounded by figures such as marshal, chancellor, counselor, or priest, which do not belong directly to the military sphere but function as advisory officials. Only then come the figures that represent "military people." Instead of a martial metaphorical framework, as prevails

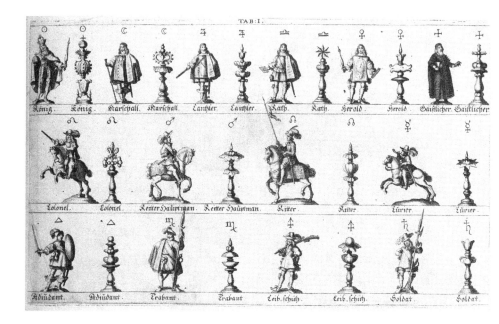

Figure 2.4
Figures and designations from Weickmann's King's Game.
Source: Herzog August Bibliothek Wolfenbüttel: A: 5.6 Pol. 2°. Reprinted with
permission.

in the work of Grimmelshausen and other baroque writers, Weickmann's
game description speaks of "insult" and "protestation."[41] If a figure that
portrays a simple soldier can take a hierarchically higher figure, then it has
to decide whether it wants to assume that figure's official post. If it declines
to do so, it might be able to take the place of a still higher figure in the course
of the game. Once, however, it has assumed the post of a figure, it is committed to that role until the end of the game. If chess has always stood for
the military confrontation among rulers, Weickmann turns the King's Game
into the symbol of the battle for the offices of a kingdom.

The title page of Weickmann's manuscript illustrates by iconological
means precisely such a power constellation: seven electors are absorbed in
Weickmann's game, and the Kaiser[42] is elevated to the level of the game
board and, as it were, put at stake (figure 2.5).

The electors are labeled with the cardinal virtues on banners. They hold
a letter, a book, or a marshal's baton, but no weapons. In contrast, armed

Figure 2.5
Title copperplate from Christoph Weickmann's "King's Game," Ulm 1664.
Source: Herzog August Bibliothek Wolfenbüttel: A: 5.6 Pol. 2°. Reprinted with permission.

warriors line the edge of the scene. In the foreground, they slay evil crea-
tures. Under the table on which the game sits, there are demons in chains.
Weickmann poses the question of power in the face of the power that
emanates from the weapons. Among all the doctrines of his writing, questions of weaponry stand out: whether "subjects should be allowed to carry
weapons,"[43] whether their rulers should "instruct and train them well and
adequately in war exercises, defense and weapons"[44] and whether "private
persons, citizens and subjects should neither be granted nor allowed to
have all too many weapons?"[45] Weickmann's game delineates the modern
state with its standing armies, its civil service and its monopoly on
violence.

It becomes increasingly decisive who speaks in the service of the king
and how he speaks. The King's Game coincides in one respect with the
core of any dispute: for "eruptions of temperament" rupture the framework
of the fictive game, insofar as affects withdraw from the register of simulation and dissimulation. For this reason, Weickmann recommended his
game for the testing of new state officials and claimed "that through this
game a high-ranking person could thus investigate and interrogate all
distinguished officials' temperaments easily and without any effort, which
cannot otherwise happen so easily."[46] To put the officials' temperaments
to the test, the game challenged its players to form alliances. The electoral
arithmetic that finds expression in the game resembles the perpetual threat
in the seventeenth century that with the appointment of an eighth elector,
an equality of votes could occur that would prevent any sovereign display
of power. Ultimately, the calling-into-question of the three clerical and four
worldly electors who elected the king—with respect to both their number
and their denominational affiliation—contributed to the outbreak of the
Thirty Years' War. Weickmann's title page therefore stands for representatives ensnared in a struggle for their own form of rule. It is probably no
accident that elector Maximilian Heinrich, archbishop of Cologne, is the
first among the addressees to whom the writing is dedicated. The King's
Game does not stage a hostile power that threatens to break in from
outside. It shows a battle that has turned inward.

Weickmann presumably modeled the arrangement of his game boards
on the designs of his friend, the Ulm architect and engineer Joseph Furttenbach. He might have also had in mind the cruciform battle formations
of the most renowned German military historian of the seventeenth
century, Johann Jacob von Wallhausen (figures 2.6 and 2.7).

Figure 2.6
Game board of Weickmann's "King's Game" 1664.
Source: Herzog August Bibliothek Wolfenbüttel: A: 5.6 Pol. 2°. Reprinted with permission.

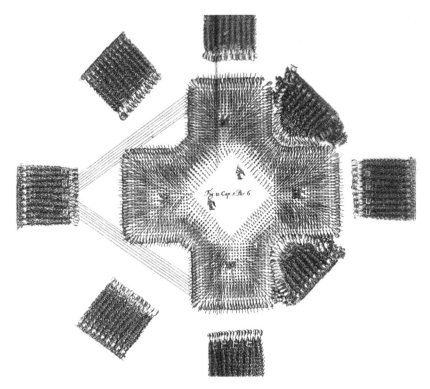

Figure 2.7
Attack on a defense formation of foot soldiers, depiction from Johann Jacob von Wallhausen: "Art of War for the Infantry," Oppenheim 1615.
Source: Wallhausen, J.J., Kriegskunst zu Fuß, Faksimile, ARA7 1971, reprinted with permission.

Whether it was the star-shaped formations of redoubts designed by Simon Stevin or fortifications designed by Furttenbach, they are directed outward in expectation of the enemy and its forces. In Weickmann's game as well as on his programmatic title page, all the forces revolve around a center that lies at the heart of the star-shaped construction. Furttenbach published a noteworthy design based on the same octagonal layout that identifies four chambers as stages. In the center of the construction is a table designated for twelve people that can be aligned with the stages through a turning device. The stage sets are similarly conceived as mobile, so that one can speak of a double multiperspectivism (figure 2.8).

If Weickmann, with his game, develops a topology and a set of rules that endanger the power of the one through the polyphony of the players,

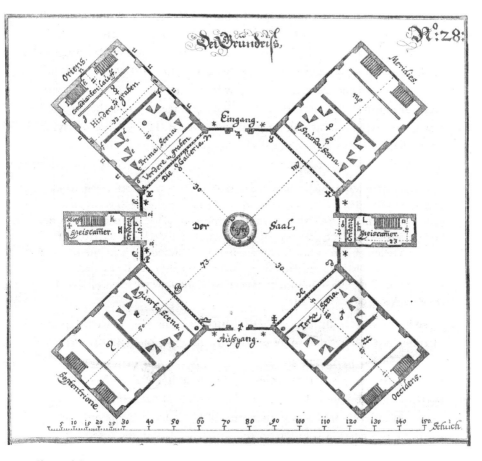

Figure 2.8
Layout of the "Theater Hall" by Joseph Furttenbach with rotating table in the middle
and four stages, Augsburg 1663.
Source: SLUB 23.4.656. Reprinted with permission.

Furttenbach ties the gaze of the potentate to a stage machinery that frag-
ments the world theater into multiple stages.

Play as a Bastion of Knowledge

Whether Leibniz, with his curiosity about games and instruments, was also
acquainted with Weickmann's game is uncertain, but quite possible. Naturally, a copy of the book is available at the library in Wolfenbüttel where
Leibniz was appointed librarian; after all, the book is dedicated to the

founder of the library, Duke August. Leibniz's secretary Joachim Friedrich Feller imparts that Leibniz had spoken of a King's Game "where the prince, chosen by lot, gives orders."[47] Weickmann did in fact propose drawing lots to decide which players would compete against and with each other. But even if Leibniz had been acquainted with it, his own war game designs went in another direction. He noted that "arrangements of the depicted war game" would allow—along with fortification models—the replaying of lost battles.[48] In his thoughts on a "German military system," he elaborated further on the proposal:

Newly invented war game, military colonels and captains, also other commanders practice it instead of the chessboard and card game, and come to greater science, speed and *invention*; one could represent with certain game pieces certain battles and skirmishes, also the position of the weapons and the lay of the land, both at one's discretion and from history, for example if one wanted to play the Battle of Lützen, the skirmish with the French at Ensisheim and other such historical events; thereby one would often find what others missed and how we could gain wisdom from the losses of our forerunners.[49]

Leibniz—who, with his theodicy, opened up a space for the conception of other possible worlds so as to identify the best of them—is also the inventor of counterfactual military historiography. But he did not stop at the idea of reenacting past battles in the game. If Weickmann could not do without elaborate color and number coding of the game boards and figures in order to orchestrate the events of the game, Leibniz proposes in his military system a solution to the problem of how dispersed soldiers "can assemble themselves in battle, if the regiments differentiate themselves with colors, the companies with the strokes or lineaments of the colors or numbers. Thus everyone can recognize from afar his regiment and [from up close his] company."[50]

Leibniz not only poses the question of the correct formation of sign systems, but also that of how other fields—including battlefields[51]—can take on formations from sign systems.

The alignment with games is the key to managing areas of life that elude Leibniz's program of rigid calculability. Though his publications do not reveal it, he systematically analyzed diverse games; he was among the first to examine the correspondence between Blaise Pascal and Pierre de Fermat on games of chance—he opposed it with his own probabilistic calculus.[52] He urged Jakob Bernoulli in an exchange of letters to publish his

ars conjectandi, which formulated the law of large numbers. And he investigated newly emerging games such as solitaire.[53] Leibniz's designs for an
academy of games and his "Drôle de Pensée" are places where he deals
explicitly with games. Usually, however, he pursued his game analyses in
secret and in all seriousness.

After many false starts, Leibniz ultimately managed to establish an
academy in Berlin—not least of all because he proposed financing it with
profits from a still-to-be-created lottery monopoly.[54] In the first issue of its
magazine, with the programmatic title "Berlin Association for the Promotion of the Sciences," Leibniz begins with an epistemology of games: In
them, he argues, people are more inventive than anywhere else. The mathematics of games does not deserve attention due to the object itself, but
rather with respect to the *ars inveniendi*.[55] What games of chance achieve
for mathematics has been demonstrated by Blaise Pascal, Christian
Huygens, and Pierre de Fermat with their calculations of probability. But
games that combine chance and skill are capable of far more. They provide
the best representations of human life, especially in military affairs
and in medical practice, which rely in part on skills and in part on
contingencies.[56]

Leibniz exemplifies his program with his own analyses of the game of
solitaire and in his invention of a game that simulates ship maneuvers.
Finally, he cites an illustration that shows Asians playing a game that we
know today as "Go." The game, according to Leibniz, relies on skill alone
and not on chance, and it is played in China mostly by senior state officials
for whole days. Here the game pieces are not taken, but surrounded. The
winner is the one who takes the freedom of movement from the other:
"so to speak, without murder and blood. Though this is not uncommon
in other games, it is compulsory here. . . . [It] is known that the peoples
of Southeast Asia behave in this matter in, so to speak, a more Christian
fashion than those who call themselves Christians, and as a rule avoid
killing specifically in war."[57]

At the end of a century that threatened to be submerged by the devasta-
tion of its sectarian civil wars, at the end of a scholarly life that discovered
new worlds in the mere unfolding of its signs and semiotic operations, and
at the beginning of a mathematical influence poised to free itself from its
magical and mystical roots, this late publication—which is followed by an
article on his calculating machine—reveals in a condensed fashion a desire

that seeks to read in the play of semiotic operations at once the most immediately evident and the immeasurably distant.

It should be recalled in conclusion that Martin Heidegger linked Being as grounding without ground to the word and the object of the *calculi*, insofar as that can mean calculating stones as much as game pieces. "When God calculates, the world comes to be"[58] is how he translated Leibniz's "*Cum Deus calculat fit mundus*," only to offer still another reading: "While God plays, the world comes to be."[59]

3 The State of the War Game

The War Game

In light of the devastating consequences of the Thirty Years' War, Christoph Weickmann—with his King's Game for the determination of "distinguished officials' temperaments"—evidently pursued the goal of recommending to the potentates of his time the consolidation of a professional class as much as a means to their rise. Had his work had a broader reception than was actually the case, he would most likely have himself become—not completely un-self-servingly—the prototype of the very official advisor and administrator in military affairs to whom he assigned a decisive role in his game. In actuality, however, another half-century would elapse before the "soldier king" Friedrich Wilhelm I, in 1713, after the War of the Spanish Succession, came to the realization that it was not enough to keep soldiers permanently in position. To ensure the maintenance of a standing army demanded first and foremost officials with cameralistic skills. He recruited them from among the officers of his army and thereby opened up the possibility for them to switch from a purely military career to an administrative one. The offices—those chancelleries that the Holy Roman Empire created for the administration of its provinces—were now increasingly open to officers who had defied all literacy campaigns for centuries.

Equipped with the highest official status, they took up their posts in the *General-Ober-Finanz-Kriegs und Domänen-Direktorium* (known as the *General-Direktorium* for short) and in the numerous war and domain chambers of the provinces. War contributions and tax revenues now flowed together under one umbrella. Plans for the supplying of the armies and the precise elaboration of deployment plans were managed on site by the war and domain chambers. The administrative structures of the aspiring

Prussian power thereby countered the borders of the German regionalism and created with their war councils an alliance between officialdom and the military. Though the soldier king scarcely got involved in any larger battle in his lifetime, the dimensions of his battle plans alone forced one to look beyond existing borders. Moreover, Prussia's strategic planning work would extend over generations of Hohenzollern kings. Initiated by Friedrich Wilhelm I, the elaboration of battle plans was continued through Friedrich II and III. But ultimately, the strategic designs for various war theaters encountered a limit. This limit neither resulted from insurmountable natural conditions nor was dictated by superior hegemonic powers. The absolute limit of strategic cabinet wars turned out to be the incalculability of tactical space.

Attempts to subsume tactics as a special case of strategy fail on all levels. On the scale of the strategic, the particular does not appear, but rather vanishes as a negligible quantity in the balance of forces. These are delineated in the first population surveys of Johann Peter Süssmilch, reflected in the trade balances of prosperous provinces and embodied in the conscription of ever-larger armies and the recruitment of mercenary armies. The invisible hand taken into account by Adam Smith appears all the more transparent the more effectively the war and domain chambers succeed in revealing the productivity of the body of the people in their documents and orchestrating it by administrative means.

The tactical space of the battlefield, however, eludes a cameralistic order: in tactical space, events obey entirely different temporal constituents. Events are beholden only to the moment and transform space into an operational field of visibilities and invisibilities, which refuses any retroactive representability.

Three remarkable individuals whose paths cross in Berlin shortly before the wars of liberation show how the Prussian military power was inevitably brought into confrontation with tactical space. The first is Carl von Clausewitz, who developed his theory of the small war in 1810–1811 at the General War Academy. At the same time, Clausewitz provided military education to the princely sons in the court of the Hollenzollerns until 1812, when he switched sides and joined the Russian services. The void that he left behind in the Prussian court was filled in tactical questions by the war counselor Baron George Leopold von Reiswitz. His career as a soldier was preordained by family tradition. A medical malpractice, however, had cost him the necessary physical integrity. As a result, Reiswitz took up the development of a war game that would cause a stir first within the Prussian Army and ultimately worldwide. His initial workplace was the war and domain chamber in Breslau.[1] His achievements there ultimately brought him to the court in Berlin.[2] A second lieutenant of the Prussian army named Heinrich von Kleist also returned to Berlin—after a less successful career than that of Reiswitz in the Königsberg war and domain chamber—and hoped after further professional failures to be able to serve the Prussian court again as a soldier. It would be Heinrich von Kleist's last attempt to render his fatherland the absolute service that he propagandized with each of his plays.

But ultimately, it was above all the war games of Baron von Reiswitz that would mobilize the armies in a hitherto completely unknown fashion. Reiswitz's son—called by contemporaries a "military Faust"[3]—played a decisive role in that mobilization. Like Kleist, he would end his own life when the appointment to a military post failed to materialize. The two chapters that follow are therefore devoted in particular to these two Prussian soldiers whose war games claimed their own lives as their first victim.

Kleist's War Games

The teichoscopies and messengers' reports in Kleist's dramas do not employ the old theatrical trick of Greek tragedies, which instead of staging great battles—and how could they?—merely legitimize their description. In contrast, Kleist's plays, such as *Die Hermannschlacht* ("Hermann's Battle"), *Die Familie Schroffenstein* ("The Schroffenstein Family"), and *Der Prinz von Homburg* ("The Prince of Homburg"), show how the transmission of bad news, declarations of war, and attack orders are by themselves capable of initiating dramas and thereby of putting the life of the messenger at stake— or it's the messenger who puts the lives of others at stake. It is less the precarious contents of the messages that provoke the dramatic twists than those incalculable moments in which the written word is rendered inoperative—for example, when the Prince of Homburg flouts a concerted battle plan and proceeds to attack on his own authority. Such disturbances of the transmission avant la lettre are generated by the shift from the medium of writing to the word; Kleist stages the affect as its most drastic communicative effect.[4] This effect cannot be captured more precisely and concisely than it is with Wolf Kittler's formulation that Kleist's dramas are without exception about the unfolding of the "function of writing" as affection.[5] Thus Kleist's plays and novellas seek to inscribe themselves in battlefields and military hearts, deeper than the toughest drill and the most distinct command ever could.

But Kleist's patriotization of hearts in the run-up to Prussia's wars of liberation has a price: a military buildup induced by poetological means will mobilize the masses all the more successfully the less it ultimately controls them. Only for Prince Friedrich von Homburg is the dream of the laurel-crowned war hero fulfilled. He flouts the royal order, recklessly endangers his life and those of his comrades-in-arms, but in the end helps his fatherland triumph and is pardoned. When Kleist wrote the play between 1809 and 1810, he might have hoped that a no less reckless action—which took place on the periphery of a battlefield at Aspern, and which he himself had to answer for—would come to an equally favorable end. A memorial stone on the Kleiner Wannsee lake, which marks the site where Kleist committed suicide, testifies to the fact that the opposite occurred.

The circumstances that drove Kleist to take his life can be traced back to a double game that he played while—in the battle between Bonaparte and Archduke Charles of Austria at Aspern—more soldiers met their death than in any previous war theater. In retrospect, the massive death toll of twentieth-century battles seems to have been presaged here. By a hair, Bonaparte at Aspern would not only have lost a decisive battle for the first time, but also offered Austria the opportunity to pursue his retreating armies and defeat them—especially if Prussia had rushed to its aid.

Kleist—who had set off for Bohemia with Friedrich Christoph Dahlmann, the historian, politician, and leader of the Göttingen Seven, so as to do everything in his power to ensure "that the Austrian war would become a German one"[6]—was drawn to Aspern in expectation of a battle. However, staying at an inn, he scarcely paid attention to the looming battle. While the troops of Archduke Charles of Austria and Napoleon collided, Kleist and Dahlmann were at the inn, absorbed in a war game that "had just . . . been much improved"[7] by his friend Ernst von Pfuel and that the three of them had often played in Bohemia.[8] It also emerges from Dahlmann's account that the Prussian major von Knesebeck was present at the same inn. The sight of Kleist as an ex-officer playing war elicited a disparaging remark from Knesebeck. Kleist only countered tersely that everything was contained in the game.[9] Only when, the next day, Kleist and Dahlmann—who were still or again absorbed in the war game—were informed by the innkeeper that the battle had reached its climax did the two of them rush to the battlefield.

The threshold on which Kleist operated at Aspern marked nothing less than a field of contingencies; its representability challenged writers and reporting officers alike. The founder of the modern General Staff system, General Gerhard von Scharnhorst, feared nothing more than a narrative element that wove a story from fragmentary news and reports of a battle that had not occurred in that way and could only lead to false conclusions with respect to future battles. Therefore, he warned against the historical representations of past wars; they were no more than a "novel bordering on probabilities."[10] Instead, Scharnhorst encouraged the systematic collection of all the records that emerge before, during, and after a campaign, however incomplete they may be. For the General Staff officers, to study the ways in which this data reveals a coherent picture of the most recent battle was the best preparation for the next military confrontation.

Since Aspern, writers whose gift is not judged by the performance with which they approach the creation of a text, but only by the result, have it hard. Even the most realistic among them, Honoré de Balzac, failed at the self-imposed task of capturing the Battle of Aspern in novelistic form at the end of his life, even though he spared himself no pains with his research and did not even neglect to speak to soldiers who had taken part in the battle and to visit the battlefields.[11] All that remains of his project are the announcement of the novel and a fragment: "The Battle. First Chapter. Gross-Aspern. On the 16th of May in the year 1808 at noon."[12] Thus it was, of all things, in the preliminary stage of a completely failed project for a novel that a crystal-clear conception emerged of what borderline-hallucinatory effects the novel would now make possible:

I tell you that "The Battle" is an impossible book. In it I will make the reader familiar with all the horrors and all the beauties of the battlefield. My battle is Essling [Aspern]. Essling with all its consequences. It shall be thus: a cool head in his armchair shall see before him the region, the lay of the land, the masses of men, the strategic events, the Danube, the bridges, shall marvel at the details and the battle as a whole, hear the artillery, take an interest in the movements of the chess-board-like formation, see everything, feel in each manifestation of the great army Napoleon, whom I will not show or whom I will allow to appear in the evening, as he crosses the Danube in a boat. No womanly face, only cannons, horses, two armies, uniforms; on the first page of the book the cannon roars, on the last it falls silent;

you will read through the smoke, and when you close the book, you should have seen it all intuitively and remember the battle as if you had taken part in it.[13]

In fact, the Prussian General Staff no longer entrusted the representation of such war panoramas to writing alone and chose shortly thereafter to couple the written data with the war game apparatus of Baron von Reiswitz. The phantasm, however, remained the same, for Reiswitz too intended his war game to serve the purpose that he found demanded in a provincial newspaper: in the future, an officer shall be spared the journey to the "four Silesian battlefields," because a war game "could conjure" them "into his room," along with "the remaining, eternally memorable battle theaters of Silesia, in order to maneuver variously with . . . figures on them."[14]

Knesebeck, however, at the Aspern inn, might have belonged to the last generation of General Staff officers who still found a war game largely absurd. To appreciate what a rapid development the war game underwent in order to ultimately become a decisive basis of military action, it is sufficient to cite a single episode. It was recorded by the "Historical Division, Headquarters, U.S. Army, Europe" in the course of a clarification of the basis on which the German armed forces were able to plan their blitzkrieg operations in the first place. For this purpose, after the end of the Second World War, the historical division had a study prepared by barracked Wehrmacht generals.[15] The report of the infantry general Rudolf Hofmann made a particularly strong impression: in the course of the Ardennes offensive, the staff of the Fifth Panzer Army held a map exercise on November 2, 1944, to defend against the attack by American armed forces. General Field Marshal Walter Model was in charge. (He had replaced the Hitler opponent and Kleist relative General Field Marshal Ewald von Kleist.) All of the key commanders and their General Staff officers had gathered in the headquarters. The map exercise had scarcely gotten underway when a report announced that the American armed forces had actually launched a counteroffensive. At that point, the assembled commanders wanted to rush to their posts, but Field Marshal Model ordered them not to leave the room and to continue the exercise. However, the map exercise was adapted as quickly as possible to the continuous reports from the front. "The situation on the front—and correspondingly in the map exercise—came to a head over the next few hours." But the chains of command between the commanders gathered for the map exercise and their General Staff officers could scarcely have been shorter, so that "after only a few minutes . . . General von Waldenburg, instead of issuing theoretical orders at the map table, issued his actual operational orders to his chief operations officers who were there with him and to his receivers of orders. The alert division was thereby set in motion in the shortest conceivable time. Chance had turned a simple map exercise into the seriousness of reality."[16]

Of course, the Prussian military intelligence from Aspern was still far removed from the real-time nature of the Wehrmacht's radio transmissions via ultra-short-wave. It was up to Knesebeck alone to take stock of the situation in the war theater of Aspern and personally report on it to Friedrich Wilhelm III.[17] Knesebeck had won the unreserved trust of the Prussian king immediately after the catastrophic defeat of the Prussians in the twin battles of Jena and Auerstedt, in which Napoleon's superior command structure also revealed the desolate and outdated constitution of the Prussian army. Immediately after the battle, Friedrich Wilhelm was wandering in an open field, exposed to the danger of being captured by Napoleon's troops, when "Major von dem Knesebeck of the General Staff" encountered him. Knesebeck knew the area from "earlier reconnaissances" and led the king to safety. "The king never forgot his service, and from that moment on he would remain faithful to him."[18] After Aspern, the king sent Knesebeck to scout out whether a favorable military alliance with Austria's imperial army would present itself. When Napoleon's defeat in the battle loomed, Knesebeck recognized the most favorable moment to deprive France of its supremacy in an alliance with Austria. In retrospect, Clausewitz judged the situation similarly and called Aspern a missed chance to take advantage of Bonaparte's disadvantageous situation.[19] But Knesebeck was prevented from traveling to Königsberg, where Friedrich Wilhelm III was staying, and persuading him personally of the necessity of entering into war. He was thwarted not by Napoleon's spies, through an act that today would be called "counterintelligence," but by, of all people, Prussia's most ardent despiser of Napoleon, who would have passed up no opportunity to defeat him. In Aspern, Kleist—who had long ago left the army and exchanged his weapon for a pen—had taken in hand two things that would be fateful for him: alongside Pfuel's war game, he had obtained two pistols.

In war, Clausewitz would teach, even the slightest contingencies sometimes have considerable consequences. That Prussia did not already enter into an alliance with Austria in 1809, which could have ended Napoleon's hegemony, was possibly due to a single bullet. For Kleist extended his war game at the inn. He loaded a pistol purchased a few days earlier and laid it—against Dahlmann's protest—on the table. It remained there overnight. The next morning an adjutant of Knesebeck's grabbed one of the pistols in jest and pulled the trigger. He could only just glimpse a bullet that barely missed Dahlmann's temple. But ultimately Knesebeck called out, "For God's sake, I've been hit!"[20] A summoned surgeon had to leave the bullet in Knesebeck's shoulder. Due to the gunshot wound, all that remained for Knesebeck to do was to convey his situation report to Friedrich Wilhelm through a messenger, knowing that his words would lose their urgency in Königsberg. When, after weeks and repeated correspondence, Friedrich Wilhelm ordered Knesebeck to promise Austria full military support, Napoleon had already sealed the pact with Austria through his marriage to Marie-Louise von Habsburg. Knesebeck's biography concludes with the words "The Prussian patriots were cheated of a new hope."[21] This source on soldierly leadership does not mention that it was Prussia's probably most patriotic writer, of all people, who played the decisive role in Knesebeck's gunshot wound. Thus, scholars today can largely perpetuate the legend that Kleist foundered with each of his undertakings on a state and a society that were not yet ready for his modes of life. Perhaps now is the time to ask, conversely, how far the Prussian reformers' experiments and readiness to take risks went.

When, in 1811, it seemed to those Prussian reform forces—with the military officer Gneisenau and the statesman Stein at the forefront—that an alliance was possible, this time with the Russian Tsar, which augured a promising war against Napoleon's rule in Prussia and his supremacy in Europe, Kleist too held out renewed hope for a military post. Every position and every task that the Prussian Junker had set himself thus far in order to secure his writing had ended in a fiasco: he had failed as a Swiss farmer, was taken prisoner as a spy by the French for half a year when he was working for the Königsberg civil service, and was ruined financially as a magazine and newspaper publisher. Perhaps he was now hoping to attain himself what he had bestowed on the Prince of Homburg as a plot: the awakening from a dream into a reality that turns out to be a nightmare, but in the end still provides the twist, the fulfillment of the longed-for dream. After the incident during the battle at Aspern, a considerable amount of diplomacy and knowledge of the most recent military practices would have been necessary for Kleist's reinstatement in the Prussian army. Almost no one but his closest confidante and cousin Marie von Kleist could have managed to accomplish this feat. First, she sent him to

General Gneisenau with "military essays." Second, she recommended him to the king for his royal guard.[22] Previously, she had written about Kleist, "My gracious kind king should not believe that his youthful adventures, his poetic peculiarities are unknown to me, all these things have elevated and augmented his sense of patriotism, only enthusiastic people will now amount to anything."[23]

At this moment, Kleist's adventurous life and the outgrowths of his poetic "peculiarities" seem to coincide; probably it is otherwise only in his dramas that one should expect that the person recommended to His Majesty as a member of the royal guard is the very tragic figure who had previously recklessly put in jeopardy the life of another man—someone who was responsible for having saved the king from a dangerous situation and to whom the monarch had subsequently entrusted his life. Marie von Kleist, however, not only asked for clemency for Kleist's past transgressions, but also cited his merits: "For several years he has also occupied himself a great deal with tactics. Played war games, etc. etc."[24] Kleist delivered the letter to Friedrich Wilhelm III in person—in an audience that the king had granted him.[25] It was probably that very same day that Friedrich Wilhelm III issued an order that promised him a military post in view of the approaching war against Bonaparte. But only shortly thereafter, the king chose an alliance with Napoleon and thwarted the insurrection plans of the Prussian reformers around Baron vom Stein. Gneisenau, Grolmann, and Clausewitz changed fronts and volunteered for the Russian, Austrian, and Spanish armies. To serve as an officer under various military leaders had been a common practice for centuries. From that point on, not identifying with one's native army no longer meant *not* fighting for the fatherland. On the contrary, to fight for the fatherland meant, above all, recognizing an absolute enemy.

Shortly before Kleist had traveled to Aspern, he met Stein in Austria along with other reformers; he shared with Clausewitz friends of similar sentiments and also a table.[26] That had probably been reason enough for Gneisenau to receive him for extensive discussions. But when Friedrich Wilhelm III avoided a military confrontation with Napoleon, Kleist— unlike the most radical Prussian reformers—did not even have the possibility of changing fronts. On the Kleiner Wannsee lake he once again loaded two pistols—this time for himself, weary of life, and for the dying Henriette Vogel.

Clausewitz's "Factory of Tactics"

At the end of the First World War, all General Ludendorff could do after a final failed offensive was bemoan publicly the failure of policy and state privately that his "strategy was defeated by the dominant tactics."[27] But the significance of tactics had already come to the attention of strategists a century earlier; thus, Gerhard von Scharnhorst had recommended to Clausewitz the Mark Brandenburg for the study of the small war because of the nature of its terrain. Clausewitz subsequently noticed that what "is foreign to the large war, 'observation of the enemy,'" was peculiar to the small war.[28] In contrast to the battalions of the large armies, which were issued daily marching orders and one-time attack orders, the free-floating and light troop units of the small war did not receive orders but "missions."[29] The signal to attack was derived from continuous observation and news of the terrain and the enemy. The fact that one should always be suspicious of the information belongs, as Clausewitz asserts, "to that wisdom to which, for want of anything, better scribblers of systems and compendia resort when they run out of ideas."[30]

The small war would expand considerably. Ultimately, the tactical insights that were gained from it began to develop into the predominant forms of apprehending the battlefields. From that point on, it seemed imperative to enlist officers with the independent faculty of judgment. The Kantian philosophy of the enlightened subject therefore found an early ally in the doctrine of the reconnaissance soldier of the Prussian army—indeed, the German *Aufklärung* tellingly signifies enlightenment as well as military reconnaissance.

Perhaps the most memorable formulation of what enlightenment is can be found in one of Immanuel Kant's footnotes:

"*Thinking for oneself* means seeking the supreme touchstone of truth in oneself (i.e., in one's own reason); and the maxim of always thinking for oneself is enlightenment."[31]

Kant's definition of enlightenment is part of a text that was directed programmatically toward a broader public: "What Is Orientation in Thinking?" The concept of orientation, as Kant argues there, should be understood literally, which means, first of all, geographically:

"In the proper meaning of the word, to orient oneself means to use a given direction (when we divide the horizon into four of them) in order to find the others—literally, to find the sunrise. Now if I see the sun in the sky and know midday, then I know how to find south, west, north, and east."[32]

Accordingly, Kant develops the condition of possibility for a concept of orientation that first stems from a geographically empirical datum, then permits conclusions from "accidental perception through the senses"[33] and ultimately delineates positions on the basis of the "pure concept of the understanding."[34] The scenarios by means of which Kant exemplifies these three forms of orientation also make visible, albeit only implicitly, a cultural-technical development. The subject knows how to accomplish geographical orientation through the differentiation of natural givens—for example, through the determination of the position of the sun. Mathematical orientation, however, is also possible "in the darkness" of a closed room because the room and the objects in it are constructed—and thus easily navigated—spaces and bodies. But only with a reason based on subjective grounds, which is permitted to presuppose and assume "something which reason may not presume to know through objective grounds,"[35] is a mode of thought introduced for which it becomes necessary to orient itself "in the immeasurable space of the supersensible, which for us is filled with dark [thick] night."[36] Kant tests the concept of enlightenment as the task of orientation against the background of a literally understood darkness. Indeed, he initially uses the concept of enlightenment and that of darkness not in a metaphorical sense, but lets access to space by means of the understanding enter into competition with the possibility of the direct perception of physically illuminated spaces. Only with the last step of his argument is the concept of the "space of the supersensible, which for us is filled with dark night" to be understood as an analog. Here it is no longer physical and metaphysical modes of orientation that compete for access to spaces. Instead, spaces are determined from the outset by the circumstance that they are inaccessible to the senses—indeed, are "supersensible." The consequences that Kant's metaphysics of a nature that is ultimately inaccessible to the senses brought in its wake need not be the concern of this study. Decisive for the question of new forms of apprehending war is the coincidence of Kant's transcendental philosophy with the creation of a schema of the enemy that no longer exerts its power through the conspicuous presence of signs of dominance, but if possible conceals itself in space, in order thus to become a manifestation that can potentially appear everywhere. A hostile nature and an enemy attuned to nature demand a military subject who must orient himself first and foremost in thinking and is thrown back entirely on his reason, while there is no longer any basis for trust in supposedly objective grounds.[37]

Only this disposition of thinking explains the fact that the "enhanced order to think for oneself," addressed to Prussia's universities by Minister von Fürst, initially and above all encompassed the military education system.[38] That was brought about not least of all by Kant's pupil Johann Kiesewetter, whose lectures at the Berlin Pépinière academy were heard by Clausewitz, among others.[39] Prussia's war of liberation began not only with a secretly planned military reform, but also with an openly waged educational offensive directed equally at prospective civil servants and officers. Thus Clausewitz lamented to Gneisenau when in 1810 not only the Berlin University but at the same time the General War Academy opened its doors:

"Half against my will I have become a professor; together with Tiedemann I am to teach tactics at the new War Academy for Officers. In addition, I am instructing the crown prince—as you see, my occupations are nearly as peaceful as planting cabbage."[40]

The "Tiedemann-Clausewitzian factory of tactics"[41] taught the small war—which provided "a useful introduction to the modern art of war as such"[42]—to the thirteen- and fifteen-year-old princes at the court and the officer candidates at the War Academy. In the lectures on the theory of war that Clausewitz gave to the crown prince, he developed probability and friction into key concepts of his theory. Two decades would elapse before a period of peace allowed him to summarize his theories in the work *On War*, even though his book remained unfinished due to his sudden death of cholera.

Even before the wars of liberation, Clausewitz had acquainted the future officers at the War Academy with the role of a new type of soldier, whose "enterprising spirit" corresponds to the "hussar and Jäger," and who has to adapt himself in the small war to a "free play of the intelligence . . . this clever union of boldness with caution. . . ."[43] Before his royal pupils, the professor cannot help confessing that the available means of illustrating war scarcely suffice:

"The whole conduct of war resembles the working of a complex machine with immense friction, so that combinations that are easily designed on paper can be carried out only with great effort."[44] Clausewitz's war machine goes beyond Newtonian mechanics and algebraic systems of equations—on which Leibniz's art of war still relied. It was no longer tenable to neutralize incalculable friction losses from the outset through idealizations. Moreover, Clausewitz's war machine is not only based on the physical domain, but equally depends on psychic efforts and a fighting spirit that had to begin where the legitimacy of mechanics was no longer assured.

Although Clausewitz captured the crown prince's attention for his war theories only with difficulty, his younger brother Wilhelm and his Dutch cousin showed their enthusiasm.[45] In 1811 they joined Clausewitz's lessons;[46] previously, they had been instructed by Captain Ludwig von Reiche in "the art of fortification, surveying and military drawing."[47]

Reiche had brought in another instructor: war counselor Baron von Reiswitz.[48] Unlike Clausewitz, Reiswitz did not even try to develop plans on paper that could have responded to Napoleon's tactics. Reiswitz turned to the sandbox to show the prince how one could best confront Napoleon's many small—and thereby very mobile—troop units. The foot soldiers of the revolutionary armies, levied in mass conscription, might have initially been as poorly equipped as they were unpracticed in battle formations. But Napoleon's infantry, which ultimately emerged from them, was—due to its light armaments and the mobility of its small units—one thing above all: far more incalculable than Prussia's regular troop formations. In his critical confrontation with Heinrich von Bülow's military doctrines, Clausewitz had argued that the event of battle determined by Napoleon's units could be negotiated with geometric methods no better than it could with the idea that all tactics take place in the presence of the enemy, while strategy would stand for the logical measures beyond the immediate battlefields.[49]

But with his sandbox, Reiswitz first provided a medium that made it possible to deal operationally and performatively with incalculabilities, instead of expelling them from the drill ground. Unlike other war games of his time, Reiswitz's war game does not get bogged down in temporally and spatially large-scale, strategic measures. Rather, it limits its methods solely to the tactical level, which extends between the beginning and end of a mission within a battle.

Order Out of Order: Reiswitz's Tactical War Game

With his war game, Reiswitz immediately aroused the interest of Prince Wilhelm—who, as Kaiser and commander-in-chief, would listen to Chief of the General Staff Helmuth von Moltke, who was himself among the early players of war games.[50] With his enthusiasm for Reiswitz's game, Wilhelm persuaded his father to grant him an audience.[51] The fact that Marie von Kleist praised the tactical and war-game skills of her cousin Heinrich in 1811, at exactly the same time, shows how well-informed she must have been about the military educational practices at the Prussian court.

Reiswitz, however, wanted under no circumstances to present his war game to the king in the form of a sandbox. Instead, he would "immediately have a terrain made from more solid material and lay that at the king's feet. This happened only in the course of the year 1812; the king had almost forgotten about it and was not a little astonished, after such a long time, to see displayed what was, in its form, a huge chest of drawers."[52] (See figures 3.1–3.4.)

Friedrich Wilhelm had Reiswitz's war game brought at once to the Potsdam castle, where reports from the French invasion of Russia soon reached him.[53] In exhaustive war games based on the reports, he reenacted with his sons, officers and adjutants the war theater and campaigns leading up to the wars of liberation. In the process, "the usually scheduled hour for the separation of the royal family" was often "far exceeded."[54]

Figure 3.1
Leopold George von Reiswitz's tactical war game of 1812.
Source: © Stiftung Preußischer Schlösser und Gärten Berlin-Brandenburg. Reprinted with permission.

Figure 3.2
Drawers with game elements for Reiswitz's tactical war game.
Source: © Stiftung Preußischer Schlösser und Gärten Berlin-Brandenburg. Reprinted with permission.

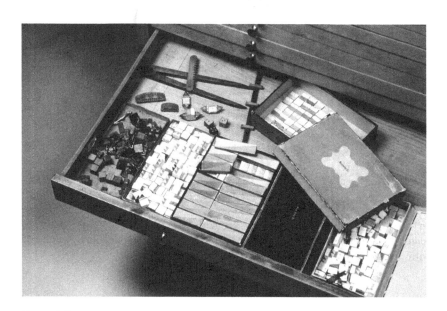

Figure 3.3
Drawers with game supplies for Reiswitz's tactical war game.
Source: © Stiftung Preußischer Schlösser und Gärten Berlin-Brandenburg. Reprinted with permission.

Figure 3.4
Detailed view of Reiswitz's tactical war game.
Source: © Stiftung Preußischer Schlösser und Gärten Berlin-Brandenburg. Reprinted with permission.

Before Reiswitz, there had been attempts by others at war game designs, such as those of the mathematician and natural scientist Johann Christian Ludwig Hellwig, who served at the Braunschweig court. In 1780, he published the "Attempt at a Tactical Game Based on Chess."[55] In 1797, the Braunschweig engineer officer and military writer Georg Venturini published the text "Description and Rules of a New War Game, for Use and Pleasure."[56] War games of this type were dismissed by the military as products of pure book learning and were granted, at best, entertainment value in the officers' casino. The fact that Reiswitz's tactical war game experienced an entirely different reception therefore cannot be explained simply by his epoch.

Reiswitz understood his game as a "mechanical device to represent tactical maneuvers to the senses."[57] But the title of his work already indicated the division of two rule systems. None of his war game rules prescribed tactical principles, whereas his predecessors had attached importance to doing just that—their games suggested that the courses of battles followed the determinism of outdated cabinet wars. In order to "represent tactical maneuvers to the senses," Reiswitz's "mechanical device" regulates the representation of visibilities, information flows, movements, strikes, and losses of troops during a battle. His rule system is thereby open to the contingencies that different tactical maneuvers can produce. Moreover, the availability of tactical maneuvers first becomes visible through the tactical war game. It can therefore be understood as a response to the distressing confrontation with Napoleon's new tactics. Earlier war games essentially only reproduced the rehearsal of specific formations. Reiswitz's tactical war game, on the other hand, is a system that confronts its players with incalculabilities that can no longer be rehearsed, but can only be played through. The systematic use of dice contributed to the unforeseeability and irreversibility of simulated courses of battle.

In this respect, too, it is evident that Reiswitz and Clausewitz sought answers to the same questions. Clausewitz, who had enjoyed a mathematical education and also strongly recommended mathematics as a subject at the War Academy, nonetheless rejected mechanical conceptions of war. The mechanics of his time still had to manage without a mathematical concept of probability. But for Clausewitz, war resembled less a mechanical system than a card game, in which incalculabilities arise from the mixing of the cards and from the opponents' unanticipatable ways of playing.[58]

In terms of intellectual history, Clausewitz is regarded as a precursor to nonlinear systems, from thermodynamics to chaos theory. Even if structural homologies with later theories in other domains are noteworthy,[59] it must nonetheless be recalled that Clausewitz claimed the fundamental absence of laws and calculability for war alone. This claim did not exactly invite theoretical transferences.

In this respect, Reiswitz's apparatus proves to be more viable than Clausewitz's analogies. He does not require compliance with tactical prescriptions. Instead, he provides the military standards for how spaces and time can be read in general. And these standards would subsequently no longer be limited to application on battlefields, but would set the course for the information and communications channels of the Prussian lands.

Reiswitz's apparatus appears primarily to be an attack on a purely incalculable and impenetrable nature, which still visibly bears Romantic traits and to which a strict framework is nonetheless ascribed. He thereby follows Kant's enlightenment program of ascribing to geographic space a mathematical and logical foundation. Nor is it an accident that the war game coincides with contemporaneous efforts to chart a comprehensive map of German provinces (figure 3.5).

In Reiswitz's tactical war game, game pieces only appear on the table once they are discovered as enemy positions through reconnaissance (*Aufklärung*) measures in the course of the game. Every game piece is thus the triumph of one's own enlightenment (*Aufklärung*) in a perceptual world that has begun to camouflage itself. In this way, the war game assists the understanding and combats the invisible enemy by giving it a form: "The condition of the enemy is invisible, one's own is before one's eyes; hence, the latter has a stronger effect on ordinary people than the former, because among ordinary people sense-impressions are stronger than the language of reason."[60]

Until Napoleon, it was still the case that the last written orders and directives were sent to the individual commanders the evening before the battle. In Reiswitz's tactical war game, on the other hand, the most important rule is not to speak, but to exchange messages within one's own ranks—which as a rule are made up of multiple players in various military positions—only via slates. Thus it is ensured that the opposing parties do not overhear what their opponents communicate and command, even if they stand directly opposite each other at Reiswitz's apparatus. Undoubtedly, it is no accident that from this point on, the written issuing of orders begins to play a role on the battlefields, which had hitherto known the use of paper only in the form of cartridge cases. In the war game, commands are not only treated discreetly through the written form in order to protect them from the ears of the opponents, but time and space undergo—much more fundamentally and in a technical sense—a discretization that leaves behind the implicit analog world view.

Impassabilities caused by morasses are reproduced in the war game as much as friction losses in communication. Whether it is a message or a troop unit that gets bogged down is equivalent and subject to the same dictate of time. All moves have to take into account a temporal standard of two minutes—that is, they represent what can be said and done in a minute in battle.[61] Under battle conditions, the artillery as a rule needed two minutes for the loading and firing of a cannon. Thus, the time window for the game moves is derived from the firepower of the heavy artillery, which dominated everything in a two-minute rhythm.

The Reiswitzian war game breaks with the previously common models that proceed from analogical movements and vary spatial and temporal processes on a scale. Instead, it retains only the effects that could also be achieved on the battlefield after two minutes on the basis of empirically established figures. Reiswitz's war game does not so much appeal to the imagination as it operates within a symbolic framework that is not even rendered inoperative when the commanders of friend and enemy stand opposite each other in the same room. The movements of individual troop pieces obey the order of the apparatus, not that of the battlefield. However the troop pieces are moved and in whatever sequence, their end position must conform to decisive characteristics of a battle that has progressed another two minutes. If a party takes less than two minutes for its moves, then the war game operates in time-lapse; if it takes longer, then the event appears as if in slow motion. Though the two parties take turns with the execution of their moves, they simulate in this succession processes running in parallel. For this, our digital existence knows the term "pseudoparallelism." The principle of simply organizing separately and according to the most efficient schemata events that do not depend on each other and do not influence each other is implemented by today's computer architectures as "out-of-order execution." Reiswitz's war game anticipated this principle.

If a troop leader encounters enemy troops in the course of the game and adopts specific measures, he conveys these only to the referee, the so-called confidant. The referee then assesses the duration in moves that accrues for the communication of the message to the commander-in-chief:

When the specific number of moves has elapsed, the commander-in-chief receives the report on the movement of the enemy, and what has been ordered subsequently by the nearest troop leader, and must, before he appears on the scene, dictate to the confidant what he wants to order.—At that point a clock is taken in hand in order to see how much time was necessary for the communication of the report, the making of the decision and the issuing of the command. Half as many moves as minutes have elapsed are brought into account and then added to the number of moves that are required to deliver the issued orders to the troop units. Only then, once the moves have elapsed, are they conveyed to the affected players.[62]

Because written orders require more time for their impartation than oral commands—as are still customary on the battlefield—their duration of communication is simply halved, and circumstances are thus reproduced in the war game that correlate to those of the battlefield. Probably at no previous time had the costs of communication been so precisely measured with the aid of a clock.

Construction and destruction are closely interrelated in the tactical war game, which provides not only miniaturized bridges and buildings made of wood and stone, but also data regarding the expenditure of time for their destruction. The referee of the war game is responsible for extensive data collection. He has to keep a record of the visibility status of troop units as much as the accrued losses. They are recorded in fractions and add up to one in the case of the total loss of a troop.

Because it was known that firearms scatter more under battle conditions than on the firing range or in maneuvers, dice come into play. They make possible chance deviations from the standards of Reiswitz's set of rules. Moreover, before the start of play, chance rolls of the dice already decimate the divisions of the two parties, so that it is uncertain with how many and which troop pieces the opponents are operating.

The playing field is formed with terrain pieces that Reiswitz called "types" in reliance on the principle of the typecase. Subsequent to the war game, Reiswitz developed a system for the printing of maps. Instead of falling back on time-consuming metallography for the production of maps, he designed a system comparable to lead printing with which different printed characters were designated for district capitals, fortresses, and other structures as well as main and side streets, forests, bodies of water, and other geographical features. From his technique, Reiswitz was hoping for a faster production of deployment plans.

Reiswitz was probably the first to register the existential fissure that separates commanders and receivers of orders when he analyzed the correspondence of two friends who played their war game by post. "A writing mistake, this correspondence of the two friends shows, once cost an infantryman his life, a case that might well have also occurred under other circumstances."[63] The war game provides training in the communication that binds a naked existence with a final authority through nothing but representatives. The strategically distanced dimension of the cabinet war has now been intertwined with the real dimension of the battlefields.

In 1812 Reiswitz had published a sixty-page detailed instruction manual to the "huge chest of drawers." It remained incomplete, because the first military movements demanded the war counselor's time. But it is doubtful that it would ever be possible to provide unlimited information on the construction of the war game. Reiswitz published only the instruction manual and intentionally declined to offer copperplate reproductions of his construction. It was necessary to prevent unauthorized replicas.

Ultimately, in 1816, Reiswitz published only the historical section of his war game text, but no longer revealed the most recent state of affairs. At the moment his war game was given a reception by the military, he promptly declared his instruction manual of 1812 to be wastepaper. Reiswitz wanted to put his papers "without any scholarly ostentation" in the hands of those "who would use them purely for an actual military purpose."[64] Consistent with this goal, Reiswitz left everything else to his son, who was about to be promoted to a second lieutenant of the Prussian guard artillery.

The War Game as War Academy

After Prince Wilhelm had, for testing purposes, assumed command in Lieutenant Reiswitz's war game as well, he declined this time to obtain a royal audience as he had done for Reiswitz's father. Instead, he sent Reiswitz—as if it were necessary to accommodate power relations to come—to the chief of the General Staff, a post that had been established specifically for Karl von Müffling and from which the military power of command would increasingly flow. One of Reiswitz's comrades-in-arms describes the meeting:

Upon our entrance we found the general surrounded by the officers of the Great General Staff. "Gentlemen," the general said to them, "Herr Lieutenant von Reiswitz will show us something new."—Reiswitz was undeterred by this somewhat cool reception. He calmly unfolded a war plan. Surprised, the general stated: "So your game is played on a real situation plan and not on a chess board?—Well, then arrange the rendezvous-deployment of a division with the troop signs for us." "I ask Your Excellency," replied Reiswitz, "to give the general and special idea for a maneuver for this plan and to designate two of your officers who will maneuver against each other. But I also ask that you take up in each of the two special ideas only that which one party would know of the other in reality."—The general was astonished, but proceeded to write down what was required. We were then assigned to the two commanders as troop leaders. The game began. One can well say that, as the maneuver developed more and more, the old gentleman, initially so cool, became warmer with each move, and at the conclusion cried out with enthusiasm: "This is no ordinary game, this is a war academy. I must and will recommend this to the army most warmly." He kept his word.[65]

The fact that Müffling kept his word is verified by the Prussians' official military journal, the *Militär-Wochenblatt*. Müffling had just taken over the editorship of the organ, which remained one of the most influential forums of the German military up to the Second World War. In early 1825, Müffling explained there:

The attempt has often been made to represent war in such a way that instruction and pleasant entertainment thereby arise. One has given these attempts the name of war game. However, in the execution, difficulties of many sorts came about, and between serious war and the light game a great disparity remained.—It is strange enough that until now only men of other classes than the military class occupied themselves with this invention, and as a result could never satisfy the demand of thoroughly educated officers with an imperfect imitation. Finally an officer has pursued this object over a number of years with attention, insight and persistence, and developed what his father, Counselor von Reiswitz, had begun to the point that war is represented in a simple and living manner. He who understands warfare in all its relations can safely take on the role of a leader of larger or smaller masses of troops in this game, even if he does not know it at all and never saw it played. The execution on good reproductions of real terrain and a frequent variation so that the diversity is multiplied through many new arrangements make the game still more instructive. I will gladly use all the means at my disposal to help increase the number of the available papers.

If Premier Lieutenant von Reiswitz has found a pleasant reward for his efforts in the acclaim of the prince of the royal house, the minister of war and the senior officers who have become acquainted with his war game, through the circulation and dissemination of the same he will not fail to gain the gratitude of the army.

Berlin, the 25th of February, 1824. v. Müffling.[66]

The decisive improvements that Lieutenant Reiswitz made to the tactical war game incorporated military developments that had not yet been available as such to his father. First of all, his version of the tactical war game required that the "affairs of the General Staff officer"[67] not be neglected—a demand that was generally made of officers from that point on, with the founding of the General Staff and the war academies. Second, Lieutenant Reiswitz tested—as a member of the artillery testing commission, together with Clausewitz's mentor Gerhard von Scharnhorst—the range and scattering distance of all available firearms, including foreign ones, on the Berlin firing range and incorporated the systematically collected data into the war game.[68] To recreate the scatterings in the war game, Reiswitz included, as his father had done, the use of dice. And third, he transferred the game to situation maps, that is, to topographical maps, which also—with their comparatively large scale of 1:8000—served the production of general staff maps (figure 3.5).

The combinatorial configuration of terrain pieces that the war counselor had devised as the basis of his war game apparatus had become obsolete ever since Müffling had considerably increased the supplies of maps with his surveying work. His passionate advocacy of Reiswitz's tactical war game reveals that he immediately recognized the expansion of the operational possibilities of his cartographic work.

Even before Karl von Müffling was appointed chief of the General Staff, he dominated the cartography of German lands to an extent that even encompassed the settings of *The Sorrows of Young Werther*. In contrast to all of his other literary works, Goethe's novel adheres to a topography that can be precisely retraced on a map.[69] In the guise of his presidency of a civilian state council, Müffling—through his surveying and reconfiguration of the land—not only provided the Weimar poet-prince topographically exact models for his novel, but also possibly served as the prototype for the character of the Captain in *Elective Affinities*.

Müffling had begun his career with surveying work in the Rhineland, which was able to build on Cassini's great French cartographic work. The mapping effort, under his leadership, had just captured Prussia when Lieutenant Reiswitz presented the tactical war game to him. Thus it is scarcely a surprise that Carl von Decker, who took over the leadership of the "Surveying and Drawing Bureau"[70] under Müffling after the wars of liberation and a thoroughgoing military reform, was also among the first to add to

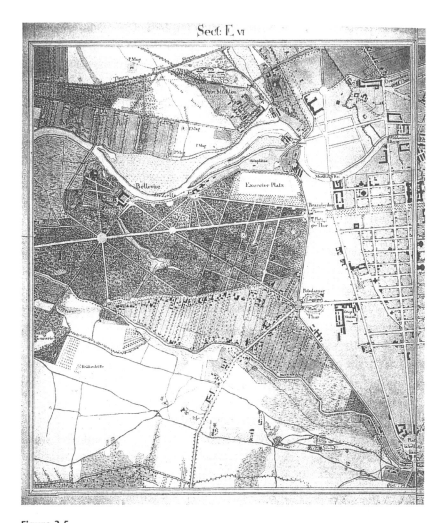

Figure 3.5
Section E/16 of war game map with a scale of 1:8000 by Lieutenant of the Infantry
Ernst Heinrich Dannhauer. The map section shows the Brandenburg Gate and the
Tiergarten.
Source: © bpk, Staatsbibliothek zu Berlin, Preußischer Kulturbesitz, Kartenabteilung,
Kartensignatur N 3660. Reprinted with permission.

Reiswitz's rules.[71] Likewise, it can be ascertained that, in General Staff education, techniques for the production of maps and their operational application in tactical war games were imparted with reference to each other.

It is striking that no institution claimed the task of regulating the development of the war game or laying down its rules, even when it was enlisted in the testing of officer candidates. Proposals for changes and additions of rules were always allotted to the unofficial section of the *Militär-Wochenblatt* and were the responsibility of authors, not institutions. Quite rapidly, societies formed in which officers devoted themselves to the war game— among them, Helmuth von Moltke, who was an "enthusiastic war-gamer"[72] from the beginning of his career and belonged to the first generation of graduates of the War Academy who were also educated in the war game.[73] That the war game not only indirectly advanced his career is apparent from his autobiography. As a destitute General Staff officer on furlough he received, through Müffling's mediation, his first military post in the Ottoman Empire only because the Turkish minister of war, Chosref Pasha, wished for an introduction to the war game. Chosref was not only interested in the Prussian military system, but in particular in the strange gift of the war game that Friedrich Wilhelm had given him. To give Chosref an understanding of the game, Moltke unfolded a map of Leipzig, "improvised a general idea and arranged a small skirmish of cavalry against infantry before a march-past and, like Squenz the role-player, more or less played the confidant of both parties at the same time."[74]

Thereafter, Chosref Pasha inquired with the Prussian regime whether Moltke could be transferred for three months to provide further lessons. As the answer was long in coming and a departing ship compelled the decision for or against the journey, the rhythm of the war game seems to have been inscribed in Moltke's reasoning. In his letter home, which sought to justify the reason for what was ultimately a four-year absence, there is the succinct sentence: had to "make (?) my move (?) within the minute (?)."[75]

Georg Brandes recognized lucidly that Moltke unified the gaze of a topographer with that of a historian, who corrected inheritances of history just as the topographer did maps.[76] But other General Staff officers who advanced to the rank of general and determined the military fate of the Wilhelmine era had, like Moltke, also gone through the war academy of the war game.[77]

All the clocks, compasses, scales, and cartographic works with which they had met a hostile nature in the war game at the war academy they externalized in timetables and code books of railroads and telegraphs, thus adapting the battlefields to the conditions of the war game as a medium. The war game might have begun as a Prussian fetish. In the end—and without having fundamentally changed—it reflected very precisely the media that held together the German Empire.

Moltke was not the first to publicize the war game beyond the borders of the Kingdom of Prussia—Reiswitz himself had already done so. The future Tsar Nicholas I was his most eager pupil. The war game was also taken up in the British and French armies, though not until much later.[78] The war game was not only a gift for allies; opposing armies also took it up of their own accord. One not only played the opposing parties; the opposing parties played too. In other words, the war game thwarted friend-enemy schemata in order to fight out those very schemata.

Reiswitz, who introduced Europe's military leaders to the war game, foundered in the end on his immediate superior. Not only was a vacant post as company commander refused him, but in addition, he was transferred from the guard to the line in the provinces. Reiswitz shot himself during his first home leave. A year after his death, a supplement appeared that built on Reiswitz's war game instruction manual without a single mention of it or him. Among the innovations of the supplement were the exceptional roll of the dice and an emergency die. If an improbable exceptional roll succeeded, the emergency die decided whether the exception took effect. Because if the point was "not to exclude any case that is possible in war, even so improbable a case, the game must also permit exceptions to the rule that must, however, have their own rules in turn."[79]

After the first quarter of the twentieth century, Carl Schmitt famously coined the formula of the sovereign who asserts himself by commanding over the state of exception. But the subject of the sovereign is a war game. By the time Schmitt became aware of this, he also had to realize that he had long since become part of one.

4 Historiography in Real Time

Theater of War

After the Second World War, Carl Schmitt withdrew into the private sphere of his Sauerland home and remained confined for the last thirty-eight years of his long life to his birthplace of Plettenberg and his parental house.[1] At that time, Schmitt was known as the "crown jurist of the Third Reich,"[2] his title as a Prussian state councilor had lost its validity, he had to give up his professorship in constitutional doctrine, and he was dismissed from the civil service. He was charged with war crimes and was arrested for two years; however, he was never convicted.

Back in Plettenberg, Schmitt grappled with the politically innocuous-seeming figure of Hamlet. Before he published his studies, he first presented them at the Volkshochschule in Düsseldorf.[3]

Upon closer inspection, Schmitt's readings of Hamlet not only lead to questions of aesthetic form and genre, but also to three central problematics of his political work. First of all, he responds (only now) to Walter Benjamin, who took up Schmitt's definition of sovereignty in his work on the baroque tragic drama and sent him his book along with an emphatic letter in December 1930.[4] Even though Schmitt had a high opinion of Benjamin's book on the tragic drama, he doubted that his concept of sovereignty was also reflected in Shakespeare's drama. In his view, the insular political relations in England seem too divorced from the developments of sovereign states in continental Europe, which were first able to produce a legal "unity of place and time and action" in classical theater.[5] Second, in Schmitt's Hamlet studies, a situation comes once again to the fore in which an outdated system of rule seeks in vain to preserve its law, but must ultimately yield to emerging powers. Thus, according to Schmitt, James

Stuart—who as James I sought the succession to the throne—provides an indestructible kernel of truth for Shakespeare's Hamlet character. This does not limit the fictional potential of language and play rules, but sets play and the "serious case (*Ernstfall*)"[6] in a relationship of tension. Through the intrusion of present time into the play, this relationship enables it to become the drama of everyone.

What takes place in Shakespeare's revolutionary century is nothing less than the transformation of England into a naval power that founds its Empire with the turn to capturing the sea.[7] Third, Schmitt is interested in the medium in which the conflict between old and new powers is fought out. For Schmitt, Shakespeare's theater is not a site where the upheavals and collapses of a time in a state of emergency are merely reflected. Rather, they find their direct representation and clarification on the stage—with the use of all the techniques of high art.[8]

Particular significance in this regard should be accorded to the trick of the play within the play in Shakespeare's revenge drama:

The famous play within a play in the second act of Hamlet is . . . doubly filtered present relevance, theater of a higher, augmented potency. The reality brought onto the stage in the drama is, within the drama on the stage, once more shown on a stage. This sort of theater within theater is only possible and meaningful where the reality of present life is itself felt to be theater, theater of the first degree, and where consequently the theater itself is essentially theater of the second degree, theater within the theater of life. Only then can the double reflection arise through which the theater within the theater leads to a heightening as opposed to a dissolution of theater.[9]

The potentialization that he sees at work proceeds retroactively: Because Hamlet stages the play at the court in order to condemn the murderer of his father (and new husband of his mother), the play in general is accorded a revelatory function that is otherwise unavailable. Schmitt therefore rigorously distinguishes Shakespeare's play within the play from later doubling strategies of the nineteenth-century public theater, which renders the play as such knowable. The latter refers to a reality that withdraws from the play—for example, when the actor apparently discards his role, speaks in an aside, and reveals himself, supposedly entirely as a private person, to be an actor.[10] For Schmitt, the later doubled play of the nineteenth century has merely degenerated to constitute a reality that is worth just as little as the play in which it is portrayed. Shakespeare's play, on the other hand, aims for ultimate answers.

To help the reader understand the difference, Schmitt makes a drastic comparison:

> The murder of James's father, the marriage of the mother to the murderer, the inhibitions and weaknesses of the philosophizing and theologizing king, all that was for writers, actors and spectators as relevant to the present as, for example, the Röhm affair for a Berlin audience in 1934. Imagine that such directly current events were brought onto the stage at that time in Berlin in the presence of the prominent figures of the regime and public of the capital in a similar way to how James's fate was actually brought onto the stage from 1603–05 in London.[11]

Schmitt's use of the conditional is certainly justified. In 1934 no stage produced a scene showing, for example, how Kurt von Schleicher, the last chancellor before Hitler's seizure of power, falls victim to the shots of an SS-commando at his desk in his Neubabelsberg house during the "Röhm-Putsch."[12]

Schleicher's department as a "background advisor"[13] had been available to Schmitt during the attempt to resist the Nazi seizure of power. After the "Röhm-Putsch," however, Schmitt defended the party chairman, chancellor and Führer Adolf Hitler through a legal apologia that certainly counts among those of his writings that brought down on him the greatest hostilities.[14] It is nonetheless difficult to apply a clear friend-enemy schema to his apologia, because opinion leaders of the Nazi's paramilitary organization, the Schutzstaffel (SS), did not let Schmitt's unconditional declaration of belief in the political Führer deter them from defaming him publicly.[15] Remarkably, in the immediate postwar period, Schmitt himself sketches a portrait of Shakespeare, the author who—in a time of unclear power relations—with his drama puts his own life at stake, while at the same time seeking to protect it through the aesthetic form.[16]

Still more highly charged than the question of what representation of political events the art of the twentieth century recoils from is the one directed at the linking element of Schmitt's comparison. There was no stage play immediately after the "Röhm-Putsch." However, there was a game in the run-up to it, which took place in Schleicher's arena of power in the Ministry of the Reichswehr, and in which "the state secretary of the foreign office, Herr von Bülow" still participated as a "spectator" from among the "prominent figures of the regime."[17] This war game—like every war game—may lack a particular aesthetic form; nonetheless, it differed from customary war games in a decisive respect. This difference pertains to the very political dimension that Schmitt regards as having been brought onto the stage with *Hamlet*. To the astonishment of the Weimar government, it could only learn from the game of its military officers—as Manstein, who would prove to be Hitler's most capable general from a strategic perspective, recognized: "We had the impression that even for the gentlemen of the foreign office, to whom such a playing through of possible cases of conflict seemed to be something entirely new, its value was absolutely apparent."[18]

Manstein neglects to date the war game. He merely indicates that he designed it at the beginning of his career at the troop office for his superior Walter Adam, who was appointed chief of the troop office in 1930.[19] The previous year, Manstein had taken over the leadership of the operations section of the troop office. Among his primary tasks was, "as the organ of the chief of the army command and chief of the troop office, to manage the great war games and exercises that served the operational training of the senior commanders and the General Staff officers."[20] The war game dealt with the "case, at the time by no means to be ruled out, that from a gradually growing political tension a violent Polish strike against East Prussia or Upper Silesia would develop."[21] Manstein therefore proposed to Adam "to precede the actual war game with a preliminary political game, in which the foreign office would participate as well."[22] Adam agreed.

Manstein's proposal to grant more significance to the military-political aspect should be regarded above all as a concession to the altered internal power constellation. The highest military organs of the First World War, the General Staff—in the form of its successor institution, the troop office—and the army command increasingly had to concede powers to Schleicher's ministry.[23] Schleicher's ministry had emerged from the Wehrmacht section that had separated in turn from the T1 section of the troop office as a military-political arena. After Schleicher had risen to the office of state secretary of the ministry (and for this position retired from the Reichswehr at the rank of major general), his long-time colleague Eugen Ott took over the leadership of the Wehrmacht section.[24]

Like Manstein, Ott too, in looking back on his area of responsibility, approaches the subject of a war game of a particular sort. He too omits an exact date, but details of his description permit the conclusion that it is Manstein's war game with the "preliminary political game."[25] While

Schleicher and his mentor, Reichswehr Minister Groener, now began to wield the influence of the chief of the army command and troop office at the top level,[26] their colleagues one level down had long been cooperating in collective war games for national protection on the eastern border, the line of which was not completely settled by the Locarno treaties. In the war game, Manstein assigned the roles: "privy councilor Köpke" had

to portray the president of the League of Nations council. Two higher officials of the foreign service took on the role of the German and the Polish foreign minister respectively. The military leadership posts were occupied by General Staff officers. The development from an increasing political tension, through illegal actions by Polish gangs, to the encroachment of the Polish army and thus the beginning of the official war was played through. The director gave the parties the picture of the general situation as it escalated on a daily basis. The military leaders of both sides had to report to the director their respective proposals and measures: the "Pole" in terms of the intended aggression, the "German" regarding the preparation of an effective defense, such as the call for border protection.

At the same time, it was incumbent on the foreign ministers of both sides to write the messages to the League of Nations through which they believed themselves capable of influencing it in the direction of their state's interests. For the legation councilor Rintelen, who portrayed the Polish foreign minister, it was thus necessary to convince those in Geneva that Poland was forced solely by German provocations to intervene. His German opponent had to underscore the constantly increasing threat constituted by the Polish measures. In this, Herr von Rintelen proved to be far superior. His gift of invention regarding alleged German provocations rendered his opponent completely speechless.

Privy councilor Köpke, who had an admirable mastery of the Geneva phraseology, understood excellently how to portray the probable attitude of the League of Nations in such a case. He presented placatory answers, the prospect of the deployment of a League of Nations commission, the back and forth about its authorities—in short, everything that one would later experience in its cases in practice—only no assertive measure that would have really deterred the aggressor.[27]

For Manstein, the war game demonstrates that "another slide into an unwanted war as in 1914" should be avoided.[28] However, completely opposite intentions are revealed by the war game if one pays attention not only to the facts, but also to the "normative power of the fictional." In that case, it is striking that, regardless of all concretely enacted scenarios, the fictional frame of a border violation remains the same—and ultimately, in 1939, with the staged attack by Polish insurgents on the radio station in Gleiwitz, becomes the pretext for war.

The Second World War was supposed to begin as a radio play, with SS men in the uniform and role of Polish border policemen and supposedly shot-down insurgents as extras. In fact, the Polish uniforms came from Wehrmacht supplies,[29] the victims from concentration camps, and a nearby prison in which one of the victims had intentionally been confined. The staging included the storming of the Gleiwitz radio station. But because the station did not broadcast any program of its own but obtained its programming from the Breslau broadcasting company, they had to make do with a storm microphone for the specifically rehearsed proclamations in Polish. With a short range, this microphone made it possible to warn of local thunderstorms.[30] To the great disappointment of Reinhard Heydrich, who had been a radio officer and had now initiated the sham attack in his new role as leader of the security police and the security service, the live broadcast of the attack was not transmitted over the radio in Berlin on the evening of August 31, 1939, as planned.[31] Therefore, the announcement of war did not come until a day later in the Reichstag and in the old medium of the newspaper. With the sentence "As of 5:45 we are now returning fire," it was therefore once again possible to claim only by means of narration what the new medium of the radio was supposed to have simulated in real time.

The Real of Simulations

For a long time, media theories have burgeoned regarding how media seem to capture the world in simulations. The fact that media themselves are not based on virtualities but on realities—which even former radio officers occasionally find hard to master—is often overlooked. Thus sociologist Jean Baudrillard has elevated the simulacrum to the central concept for the description of the state of the Western world—without, however, asking about the historical development of the simulation.

For Baudrillard, simulacra have, in an escalating fashion, first taken up the play with the real, then taken the place of its appearances, and finally themselves created a basis that is no longer dependent on the real.[32] In particular, the referential system of the image has thereby undergone a transformation that leads from reflection to self-referentiality.[33] The logic of this interpretation, however, implies that the real is excluded from the simulation media and their history is effaced.

But it is precisely the development of war games that shows how much the real intrudes into the simulation media, which is the case whenever the simulation reaches its limit and suffers disturbances. The thing to be simulated is equally affected by the intrusion of the real, even if it sometimes takes on the technical forms of the simulation. In particular, effects of the real emerge at the communicative and mediatic level. In the military-political war game, a Prussian legation councilor's ability to empathize might underlie the position of the Polish foreign minister. However, the communication practices that are thereby tested could not be more real.

It is in the war game that the difference between simulation and communication comes most clearly to light. Though infantry units in the war game that do not shoot with live ammunition—or if they do, then only at targets—and tank battalions that operate, in the absence of available tanks, with mockups or entirely on paper exclude a world of real pitfalls from the outset, their signal battalions do not proceed any differently in the war game than in war: "The headquarters or the command post is set up in houses, in the open, in vehicles or tents. The signals personnel set up the required wire and radio connections so that during the exercise the entire communications operation including the messengers proceeds as in a war."[34]

To make the situation appear to the remaining battalions "as in a war," recourse could be had to an old alliance between the film industry and the military. It should be recalled that the chief of the supreme army command, Erich Ludendorff, had already founded Universal-Film AG (UFA) in 1917 for propaganda and psychological warfare. The war game, however, made use not of film's contents but of its production methods—indeed, so much so that the general of the signal corps, Praun, would use the word "film" as a synonym for "war game": "Mostly the director [of the war game] will have assembled his 'insertions' in a sort of 'screenplay,' according to which the film then proceeds."[35] "Insertions" are measures undertaken by the war game director during meticulous plans in the run-up to the game or at the same point through the deployment of intercept companies so as to confront the war game participants with exceptional situations on the one hand and optimize one's own signals intelligence on the other hand.[36] The borrowings from film thereby serve a form of psychological warfare that is not directed against opponents but subjects one's own battalions to stress tests:

The staff must be occupied constantly in a warlike form, thus to some extent put under pressure. It is therefore the main art of the director to generate realistically the numerous "frictions" that occur continuously in war. The director must have prepared these well in the form of "insertions." Among these are the abundance of reports that amass, especially during critical situations, false reports, "rumors," out-of-date orders and reports, deceptive communications, wishes of the neighbors, inquiries of higher-ranking command posts—these should rain down on the staff and force the participants day and night to make swift evaluations and discern the essential. Among the countless individual events of war are wire faults, unintelligible telegraphs and radio reports, decoding missions, prisoner interrogations, evaluation of aerial images, assessments of the enemy demanded by superior posts, supplying missions. A captured enemy map with writing in a foreign language can also perplex the responsible specialists. Such specialists, or individual commanders or a whole section of the staff can be completely or temporarily lost in the decisive moment, while the staff operation must continue as in a case of emergency (*Ernstfall*). Such losses, caused by direct hits, paratroopers or partisans, then force the remaining staff to instructive assistance.[37]

According to General Praun, it is simply no longer tenable after a quarter-century that the army be governed

by Count von Schlieffen's brilliant vision of the future from 1909 in his essay: "War in the Present": "The commander is situated farther back in a house with spacious orderly offices where wire and radio telegraphs, telephones and signal apparatuses are available, scores of cars and motorcycles equipped for the farthest journeys await orders. There, on a comfortable chair at a wide table the modern Alexander has the entire battlefield in front of him on a map, from there he telephones rousing words and there he receives the reports of the army and corps leaders, the captive balloons and the dirigibles, which observe the movements of the enemy and monitor its position along the whole line."[38]

In fact, screenplays specifically intended to provide disturbances were not even necessary in order to falter in the simulated appropriation of new infrastructures and to sink into general chaos. The fighter squadrons that were supposed to fly a feint attack during the 1937 "motor transport exercise" had to be countermanded.[39] The traffic itself made no progress. Though the Reich's first autobahns were already in existence, gas stations had not yet been invented. Thus the tank trucks stood at the end of what was probably the first traffic jam on the new autobahns.[40] The signal corps did not do much better; their equipment for the First World War proved to be completely incompatible with new components for the coming one. Only another war game, in 1939, showed progress in the mastering of new communications methods in interplay with tank battalions.[41]

Eugen Ott's War Game of the State of Exception

The real of war games was of particular significance for the testing of communication. But that implied a special relationship to temporality. The laconism of military jargon encapsulates the state of affairs: "The locality was usually 'hypothesis,' what was played was actual *time*."[42]

The war games and map exercises did not simply dissolve temporal references through a symbolic system, but allowed a temporal extension to occur that seemed to correspond to the hypothetical situation. It was precisely because war games granted time unlimited space that what was not planned could occur. The Reichswehr and Wehrmacht hoped thereby not merely to master states of exception but to let them occur in a controlled framework in the first place.

Accordingly, it would be a mistake to believe that war games served general military training alone—when, for example, the detained Wehrmacht generals after the Second World War indicated the increasing significance of war games since 1918:

In his "Fundamental Thoughts on the Reconstruction of the Wehrmacht," which the first chief of the army command after the war of 1914/18, Colonel-General von Seeckt, authored with his own hand, he wrote, among other things: "terrain and operations studies should be performed under the hypothesis of possible military situations in the west and in the east."[43]

Even if Colonel-General Hans von Seeckt had, due to the Treaty of Versailles, gone to the trouble to stress the defensive disposition of the army, this would have been of scant significance in the context of war game studies because, on the one hand, around half of all military forces always rehearse the attack in every war game, whether they are designated blue or red, as the enemy customarily is. On the other hand, war games in the Weimar era did not first develop into a political instrument when they began to incorporate political entities. Rather, only war games were able to approximate exactly the situation that was to be avoided at all costs. As negative in result as they often turned out, demands for military buildup and reinforcement of border fortifications could be made emphatically and concretely by the Reichswehr. Even Manstein in his war game came to the not very surprising conclusion that the regime would have little with which to counter a Polish attack and could not hope for interventions by the League of Nations.

Eugen Ott did not forget to stress the negative outcome of Manstein's war game in his retrospective lecture typescript, which is specifically underlined in Carl Schmitt's copy[44]—starting in 1931, he was engaged in a close, friendly exchange with Ott.[45] Manstein's negative result might have been noteworthy for Ott and Schmitt because, in a far more consequential war game, the negative finding grew to an overpowering magnitude that on December 2, 1932, deprived Papen of his chancellorship and allowed Schleicher to become his successor. The war game, initiated and directed by Ott and presented by Schleicher in the cabinet, showed that the imposition of the state of emergency in the case of considerable inner unrest would founder on expected substantial resistances. Schmitt too had nothing with which to counter the assessment of the situation as manifested in Ott's war game.[46] This assessment invalidated all the instruments of constitutional law that he had provided Ott up to that point in order to enable the presidential regime to proclaim the state of exception on the basis of constitutional law.

Schmitt did not capitulate immediately in the face of a concrete readiness to use violence that was opposed to the exercise of law. Rather, he ceded the initiative to Lieutenant-Colonel Ott on the field of fictional hypotheses that he had identified at the beginning of his career as extremely productive for jurisprudence.[47]

With Manstein's and Ott's war games, the military complex—by declaring its means of violence insufficient—prepared to conquer the very space of symbolic operations that had hitherto been solely an affair of politics.

Since Schmitt's closest colleague in the last days of Weimar, Ernst Rudolf Huber, first broke his silence in the mid-1980s (shortly after Schmitt's death) and began to speak about Schmitt's secret missions as legal adviser to the Reich regime, Ott's war game has again become a focus of research.[48] The thesis that Ott's war game was specifically arranged by Schleicher as an intrigue to overthrow Papen as chancellor of the Reich was thereby refuted,[49] while the task of properly assessing the status of the war game has not been performed to this day. Thus, the suddenness with which the war game took the political stage could appear as unprecedented to observers today as it did then.[50] Its actual participants from the Ministry of the Reichswehr, however, practiced war games with a still scarcely investigated systematics. A comparison with Manstein's war game shows the continuity with which war games were effective independently from the interplay of power constellations. Not only does Manstein's war game incorporate various entities of the apparatus of state, which is true of Ott's version to an even greater extent, but also the doomsday scenario that Manstein's war game—under the existing presuppositions and the existing means—failed to prevent reemerges in Ott's war game, and certainly its negative outcome does too: every instability within the Reich also weakens the national security, above all on the eastern border, and positively challenges Poland's armed forces to attack.[51]

Certainly Ott's war game, in comparison to Manstein's, is incomparably more critical to its time in terms of the political dimension. First and foremost, it attempts to assess the consequences of an emergency decree that the presidential cabinet of the last days of Weimar was at the same time preparing to proclaim, while the National Socialists were insisting more and more vehemently on the takeover of the chancellorship due to their majority in parliament.

During the state of emergency from 1923 to 1924, in which executive state power was transferred to Seeckt as chief of the army command, Schleicher had already taken on the planning work for the military state of exception to be imposed on the entire Reich. The concrete implementation he left to his close collaborators Eugen Ott, Erwin Planck (the son of Max Planck), and Erich Marcks (the son of the historian of the same name).[52] In 1932 Planck rises to the position of secretary of state, Marcks to the Reich press officer, and Ott to leader of Schleicher's control center, the Wehrmacht section.[53] Horst Michael, at that time senior assistant to the historian Erich Marcks and well acquainted with his son, the leader of the Reich press office, associated closely with this circle. He also attended Carl Schmitt's political science working group at the Berlin Handelshochschule.[54] It is he who brings Schmitt into contact with Schleicher's closest colleagues.[55] After the so-called Prussian coup—the dissolution of the social democratic government of Prussia in 1932—Schmitt represented the Reich government with two colleagues. Afterward, Schmitt instructed Ott on constitutional possibilities related to the emergency decree authority governed by Article 48 and directed against obstruction strategies of the National Socialist German Workers' Party (NSDAP), the Communist Party of Germany (KPD), and the Social Democratic Party of Germany (SPD)—that is, against those parties that possessed a majority of the mandates in the Prussian state parliament.[56] In the diction of Schmitt's pupil Michael, however, the constitutional advice sounds like a description of the battles of a partisan war:

The best attack is one that drives the opponent out of a covered position and helps oneself into a covered position. With Path I the opponent can dodge into an ambush. Parties outside of the Reichstag are opponents that elude the government. They work in an area where the government cannot follow them. The government itself would have to proceed to an uncovered elevation, so to speak, where it would be exposed to all shots—With Path II the opponent is to some extent sitting in a valley where its positions can be seen and bombarded, while the government remains under cover.

Path I brings the people into still greater unrest and puts more responsibility on them than they can bear. It is open dictatorship and due to insufficient cause afflicted with the odium of arbitrary power. Path II serves the people, the government leads, educates and provides a model.[57]

After Papen had neglected to enforce the emergency decrees at a moment when voter support for the National Socialists had temporarily fallen, and as the presidential regime now itself continued to lose support, Ott advised Schleicher to probe in the aforementioned war game how even under these conditions the state of exception could be imposed and maintained. On November 18, the chief of the ministerial office, Lieutenant Ferdinand von Bredow, issued the invitations to the war game in the Ministry of the Reichswehr, which would take place a week later for two full days with the expected participation of government officials, *Führerstab* officers (*Führerstabsoffizier* was the Reichwehr's new designation for *Generalstabsoffizier*, or General Staff officer, in the wake of the General Staff's dissolution), senior military lawyers, staff intelligence officers, leaders of the *Technische Nothilfe* (Technical Emergency Relief organization)—a total of about fifty people.[58] In the Ministry of the Reichswehr, all the participants are confronted with the fictive hypothesis that the right to strike would be restricted for vital occupations on November 22, at which point it would be expected that in the next two days the SPD and KPD would proclaim a general strike, which the NSDAP would threaten to join. On November 24, the Reich cabinet would then convene so as to announce on the radio the state of emergency in the entire territory of the Reich. On November 25 and 26, the game began. It was perhaps scarcely possible to comply more with the war game principle of playing "actual time." Parts of the apparatus of state now reacted to a fictional situation for which the latest Berlin transport strike served as a template. Thus, on the one hand, the "emergency decree for the state of exception" is formulated in the course of the war game with recourse to the model of the crisis year of 1926.[59] On the other hand, a week later Schmitt and Michael draft a proclamation of the Reich president for the imposition of the emergency decrees.[60] Though the legislative and executive domains of the state are reacting only to fictitious descriptions of the situation, they do so with a previously unknown focus and readiness to cooperate that goes far beyond the framework of a merely imitative simulation.

For Ott, the war game revealed the double awareness of how a state of exception could technically be realized in the first place and that its implementation threatened to fail under the existing circumstances.[61] The accomplishment of the state of exception could not be attained with the existing infrastructural means of the military districts, the border guard, the police, and the *Technische Nothilfe* for the simple reasons that for one thing, even these forces were regarded as infiltrated by supporters of the extreme left and right parties, and for another, strikes and sabotages could aim for the systematic paralysis of the supply and transportation infrastructure, such as the Hamburg port and coal mining in the Ruhr area. Moreover, plundering of explosives and weapons arsenals would bring about an equality of armaments.[62] The Reichswehr ministry methodically sought to give a shape to the enemies of internal security, and those very enemies thus loomed just as methodically in the war game. Clausewitz's psychologism—according to which the enemy, due to its invisibility in contrast to one's own visibility, fuels the imaginary in a paranoid fashion—is now replaced by a dangerous logistics, which the enemy is assumed to have at its disposal and which first gives the enemy its most threatening shape. It cannot be chalked up to mere coincidence that it is at this very time that a young mathematician named Johann von Neumann formulates a mathematical theory that always presupposes the strategically most cunning opponent. The concluding chapter will therefore examine Neumann's game theory more closely.

When Papen—to return to Ott's war game—argues in the cabinet for a course that does not exclude the possibility of a breach of the constitution in order to maintain the power of the Reich regime through emergency decrees and against the still anticipated substantial resistance, and again recommends himself as the Reich chancellor, Schleicher asks Ott to present the lessons from the war game to the cabinet. After an impressive presentation, all the ministers withdraw confidence from Papen and urge Schleicher to have himself appointed chancellor by Hindenburg.[63]

Thus, the Weimar Reich regime found in the war game a medium for the continuation of its politics. The element of war inherent in the game, however, remained completely untouched by that. As if the transformation of Schleicher's Wehrmacht section from a former department of the operations section of the troop office into a ministerial office were not significant enough, in addition all questions of constitutional law first found clarification in the Reichswehr ministry, before the ministry of the interior was made privy. The fact that Carl Schmitt's closest confidant in the Reich government, Lieutenant-Colonel Eugen Ott, was in the Reichswehr ministry prompted vigorous inquiry from Huber's listeners after his report.[64] His formalistic legal reply that executive power is ultimately under the control of the Reich president seems strained— and, incidentally, scarcely consistent with Schmitt's legal conceptions of the ruler.

Still more precarious is a form of state that, like a Klein bottle, discards the fundamental distinction between inside and outside. Its internal instabilities are seen above all in their impact on the unresolved eastern border, which through the Polish Corridor consists of only an external frontier and contains the connection to East Prussia. Thus, Ott's war game envisioned "that communists, apparently under Polish command, acquire border protection weapons stockpiles."[65] But it is precisely that vision which can be linked to demands for additional border divisions and the establishment of militias, which are officially under the control of the interior ministry and, unlike the Reichswehr, not the Reichswehr ministry.

After Papen's resignation, Ott's war game therefore finds a continuation in two variants. The one searches for ways to bring internally motivated uprisings under control. Among other things, this includes the consideration of using tear gas—which had hitherto been deployed only in war games of the army—in street fighting as well.[66] In the second variant—at the instigation of the chief of the army command Walter Adam in January 1933, when Schmitt and Ott are still discussing emergency decrees[67]—the case is played through that forty to forty-six Polish formations advance toward Germany. As things stand, the Reich army would have at its disposal for the defense twenty-seven field divisions, thirty-four weak border divisions, and three cavalry divisions. Before the Germans, after twenty-one days of mobilization, could strike back at all, East Prussia would already be nearly lost, and two Polish armies would immediately approach east of Berlin. In any case, this is the conclusion to which Adam came in a war game that simply draws the consequences from Manstein's and Ott's previous work.[68] It is noteworthy that Adam did not even wait first for the swearing-in of a new Reich chancellor who saw himself as the first soldier of his country by a Reich president, a retired general field marshal, at the end of the same month.

The inexorable continuity delineated in the series of war games seems, not least of all, to be based on the fact that the games always reckon with regime collapses and states of exception. Therefore, the dates that are provided as fateful hours in German schoolbooks should be corrected: instead of Hitler's seizure of power on January 30, 1933, the war game on November 25 and 26 should be mentioned, because it became the condition of possibility for Hitler's chancellorship. Instead of directing all the attention to September 1, 1939, the attack on Poland, the staged attack on the radio station in Gleiwitz on August 31 deserves to be highlighted as a portent of the medium of the coming war. Instead of emphasizing solely Operation Barbarossa (the attack on the Soviet Union on June 22, 1941), it would be equally worthwhile to highlight the preceding map exercise "Otto," which owed its basis to the operational design "East" and thus to none other than Schleicher's former Reich press officer and Schmitt's close acquaintance,[69] Major General Erich Marcks.[70] Instead of remembering only July 20, it would be advisable to engage with the plan "Valkyrie," which resembled Ott's war game in decisive respects: General Friedrich Olbricht, who designed the plan, shows how internal unrest brought about by an ever growing army of forced laborers within the Reich could be quelled by a reserve army. The plan, which Hitler himself signed off on,[71] was however in fact part of the coup plans against him and was supposed to ensure the assumption of command in the Reich after a successful assassination.[72] The fact that the implementation of Olbricht and Colonel Henning von Treschkow's coup plans ultimately failed might well have been due to a fundamental dilemma: the plans could at most be played through in the framework that they simulated, but for reasons of secrecy not under the sign of the intended coup d'état. Once the attempt was nonetheless made to conduct a war game in the guise of disaster prevention and tank units moved into the government quarter, Goebbels turned out to be immediately alarmed and agitated.[73] In short, the coup plans could give themselves the appearance of war games, but they could not be tested within them.

Ultimately, it is less the end of the last German Reich chancellor, presumably as a burnt corpse somewhere on the grounds of the Reich Chancellery on April 30, 1945, that is significant than it is his last days. During those days, Hitler essentially did what he had always done over the last six war years—that is, held continuous briefings. But if one is to believe his minister of armaments, Albert Speer, then the commander-in-chief of the Wehrmacht was for that very reason spared to the last from even registering the total breakdown of his army. Rather, Speer suspected "that the General Staff under General Krebs had finally abandoned giving Hitler accurate information and instead kept him busy, in a sense, with war games."[74] At all these moments, war games fought out the progress of German history in a dual fashion: they unleashed an enormous efficacy in the domain of the symbolic, yet did not intervene in the course of events to the extent that catastrophic developments could have been anticipated and avoided through simulations.

Applicatory Method

Attempts to extrapolate from history to the future have perhaps nowhere been undertaken as intensely as they were by the General Staff. Julius von Verdy du Vernois, general of the infantry, teacher at the General War Academy in Berlin and ultimately minister of war for a short time toward the end of the nineteenth century, pursued this goal within "officer training" with the "applicatory method":[75] officer candidates, according to his teaching method, had to prove themselves in situations presented by instructors who drew their descriptions from the inexhaustible arsenal of military history and elaborated them up to the moment when key decisions had to be made. The prospective officers then had to issue orders and take measures on their own, thus demonstrating their leadership qualities. Those who found it hard to imagine a war theater on the basis of descriptions alone were aided by the semiotic system of the war game.[76] The applicatory method would remain indispensable up to the end of the Second World War.

With the establishment of the Bundeswehr and the Military History Research Institute, a debate flared up over the question of whether an applicatory approach to military history should continue to exist beyond the Second World War. The historians of the Federal Republic swiftly agreed: there was nothing to learn from military history that could be of practical value.[77] At best, a contemplative engagement with history could bring increased insight. But before tempers could cool over this, historian (and Bundeswehr colonel) Hermann Heidegger intervened. That practical value can be drawn from military history "is clearly affirmed by our former adversaries: The German military-historical source material in their hands could otherwise be given back immediately."[78] Moreover, Heidegger regarded an opposition between theory and practice as outdated: "All sciences—including the humanities—have taken on a technical-practical character."[79] With reference to Ernst Jünger's *Krieg und Krieger* ("War and Warrior"), Heidegger recalls—in vicarious continuation, as it were, of a dialogue with his philosopher stepfather—that the expansion of a "gigantic work process"[80] took place with the First World War. However, he considers the time frame within which history can yield benefits to be growing ever narrower. Carl von Clausewitz still regarded as strategically relevant retrospective views of military-historical events dating back less than seventy-five years. "The time span," Heidegger points out, "in which events are still practically instructive for us is shrinking as a result of the rapid development of weapon and transportation technology."[81] Heidegger turns to the formula of the historian Hermann Heimpel in order to designate a limit: the "present is the historian's first historical source."[82] Ultimately, it must be inferred, the applicatory method enables elements of historical processes to converge with the present of computerized operations. Thus, history becomes a system that proceeds in real time.

The compendium "The Duty of the General Staff," by Bismarck's general Paul Bronsart von Schellendorf, already expects from an officer after the Franco-German war above all the mastery of a record office: the receipt of even the most surprising information does not require an original reaction, but rather the search of the archive for comparable cases in order to master empirical multiplicity through recursive procedures.[83]

Prussia's first Chief of the General Staff Müffling had previously demanded from his officers a depiction of the Seven Years' War, which appeared in three volumes under his successor. The accounting of the

Prussian battles against Napoleon up to the wars of liberation followed, and this time too the tenure of a chief of the General Staff was insufficient for the completion of the task. The military history section under the leadership of Verdy du Vernois evaluated the reports of the wars of 1864 and 1866 and published the result as an official work. Out of this emerged, with added information on the use of railroads and telegraphs, the "Instructions for Senior Troop Leaders of June 24, 1869"—punctually before the start of the Franco-German war.[84] Chief of the General Staff Moltke nonetheless still ordered even the documentation of the Franco-German war and thereby caught up to a war that he himself had waged. From that point on, no peace lasted long enough to make the last war completely historically accessible according to the records before the outbreak of a new war. This was especially so in the Weimar era when the processing of a rising number of individual operational and tactical questions inhibited the work on a complete overview.[85] Moreover, the number of volumes to be published grew and with it the editorial period. The conclusion of the official depiction of the First World War ultimately coincides with the end of the Second World War. The penultimate, thirteenth volume—"The World War 1914 to 1918"—appeared in 1943 in a small printing: "For official use only!"[86] The fourteenth volume was supposed to make it official that the operational and tactical situation[87] in October and November 1918 by no means compelled surrender. It appeared in 1956 in Koblenz.

If generals at the time of Frederick the Great still boasted of not being able to write, the officer type of the period after the wars of liberation pursues general studies. Such was the case with Captain Griesheim, who contributed to Lieutenant Reiswitz's design of the tactical war game.[88] Griesheim was drawn "to the university, to the lectures of C. Ritter, Erdmann, Hegel, A. von Humboldt, and other men of distinguished reputation and name, while at the same time the professional studies of the art of war and military history were the subject of the most diligent reading."[89]

In an obituary for Griesheim, Bismarck's future minister of war Albrecht von Roon can only state that "it was granted to his strength of character to prove itself in a sphere that was only close to the actual military activity in the more narrow sense, not within it."[90] This "sphere" was none other than a campus that encompassed within the narrowest space Reiswitz's artillery barracks, the nearby Friedrich Wilhelm University and Hegel's house on the Kupfergraben. The fact that Griesheim was more than just one of Hegel's many listeners Roon does not leave unmentioned:

When in later years Professor Gans, after Hegel's death, published his philosophical lectures, [he turned] to Griesheim's, his friend and favorite pupil's, well-organized notebooks . . . so as to complete the unfinished writings of the famous teacher. . . . But anyone who has ever heard Hegel or even only seen one of the writings he left behind will judge what it means when experts in such a case feel compelled to take refuge in the work of a military dilettante;—thus, he was in any case among the few who had correctly comprehended the so difficult and therefore mostly misunderstood philosopher.[91]

The correct comprehension was based on a well-practiced system of writing in the mode of the General Staff, with which Griesheim and his comrades-in-arms came to Hegel's aid:

When he [Hegel] heard about a good transcript of a listener, [he had] this copied, and it was taken as a basis for repeated reading, so that changes and extensions were added to it. . . . Although Hegel always held his lectures according to a notebook, the listener could deduce the corrections, insertions, etc., from the constant turning of pages back and forth, from the searching around now at the top, now at the bottom.[92]

In addition, Griesheim made a single compilation of various notes in order to preserve all facets of Hegel's performance on the history of philosophy—a procedure that was customary in the General Staff in order to derive from countless officers' diaries new directives and war depictions. Ultimately, Griesheim would himself hold lectures, even if they were on military history at the War Academy.

According to Roon's assessment, Griesheim "should preferably have found a use in the General Staff of the army."[93] But when things actually reached that point, the war ministry was given priority.

History of the Ongoing War

After the First World War, the sphere in which Griesheim's career moved in such an exemplary manner, and which according to Roon did not come into contact with the actual "military activity," is simply liquidated. With the dissolution of the Great General Staff, the "nameless spirit of the General Staff officer"—to use the words of the chief of the troop office, Hans von Seeckt, whose responsibility this was—passed into entirely different institutions. This was surely not what the victorious powers had in mind when they required in the Treaty of Versailles, which became law in Weimar: "The Great German General Staff and all other similar organizations shall be dissolved and may not be reconstituted in any form."[94] But the legend of the Great General Staff continuing to operate, hidden in the newly established troop office of the ministry of the Reichswehr, falls short. It suggests a continuity of Prussian military efficacy—however, the latter's supposed power potential would presumably turn out to be slight in comparison to the new hybrid military-political connections that the dissolution of the General Staff made possible in the first place. The fact that a ministry henceforth has at its disposal former General Staff officers, who enter the troop office in droves, and that the power of command, contingent on the constitution, no longer emanates only from the commander-in-chief, but now also requires a decision of parliament, signifies an actual caesura.[95] But that is not all there is to be said, because the division of the power of command first provides to an expanded circle of people the possibility of legally exercising the power of command.[96] The Great General Staff has not, in any case, simply been resurrected in disguise. Rather, the division of the operationally intertwined domains of the General Staff and their embedding in various organs of the executive causes a transformation that ought to have frightened the victorious powers more than its continued existence. While recruiting, training, and operational planning staffs settled into the Ministry of the Reichswehr, the records of the military history section of the General Staff were merged into the new establishment of the Reich archive, which was under the control of the Reich interior ministry. But the Reich archive does not cease, due to its more civilian-sounding renaming, to be the operational basis of the Reichswehr. On the contrary, all the records of the wars fought for reasons of foreign policy were now connected to an internal department.[97] Therefore it is scarcely possible to distinguish what came first, the permanent fear scenario of civil war during the Weimar's early and last days or the preparation for this very scenario.

    By the time Hermann Göring, in the function of the Reich minister of the interior, urged the revival of "the old military history section of the General Staff in some way as an institution of the Reichswehr"[98] and thus worked toward the renaming of the troop office as "General Staff of the Army" that ensued shortly thereafter, the former Reichswehr officer and military historian Walter Elze had already established a "military historical section in the historical seminar of the Friedrich Wilhelm University" and assumed its direction in 1931.[99] In the same historical seminar, Hans Del-brück had only a decade earlier argued with retired Major General Hans von Haeften and his archivists about the official depiction of the First World War. Ten years later, with the start of Elze's directorate, the boundaries between military historiographers and academic military historians can no longer easily be drawn. Eberhard Kessel and Werner Hahlweg, who in prominent positions would decisively shape the reorientation of the military historiography of the Federal Republic, came out of Elze's seminar, as did Felix Hartlaub, who would be entrusted with the composition of the war diary for the high command of the Wehrmacht, and on whom Elze conferred a doctorate in 1939. The reason to elaborate on Hartlaub here is that the diaries privately written by him with an authorial intention make possible in a complementary fashion the reconstruction of the function of the official war diaries.

Even as he wavered between the profession of writer and that of scholar, which seemed equally unpromising to him, Hartlaub was overtaken by conscription. He had his doctoral adviser to thank for the transfer from the front to the historical archive commission of the foreign office in Paris. Hartlaub also escaped the subsequent stationing in the embattled area of the Romanian oil fields again only due to the intervention of his academic instructor Walter Elze. Thus Hartlaub was ultimately able to write to his beloved, the immigrant daughter of a Russian general in a Jewish oppositional circle, of life in the Führer's headquarters that he only sees everything "through the medium of the records":[100]

You ask about work. Nothing can be said about its content, of course, it really revolves around the most discreet matters, the great lines and plans of the supreme command, which are treated here in the war diary. A thinking in large masses and spaces with extensive economic and political considerations—we don't dwell much on individual little units and events. . . . We . . . simply string together dense indices and summaries of the documents that come directly from the hands of the history-making men into our folders.[101]

The fact that the "military historical illumination of this war"—which is taking place at that moment—is expected from a "military historian" like him, and not "an officer equipped with General Staff training,"[102] Hartlaub registers with amazement. An "ill-fated dissertation about an old naval battle"[103] had landed him at the instigation of acquaintances in the "military history section in the historical seminar" and now also in the "military history section of the OKW [supreme command of the Wehrmacht]."

Hartlaub, whose prose of "almost poetic sensitivity"[104] his doctoral adviser Elze—a pupil of Stefan George—had already praised in his doctorate, now authored the war diary of the OKW.[105] The great reports and overviews of the events of the war arise from a collective archival work, which the ongoing war constantly feeds and at the forefront of which Hartlaub stood. However, he did not let a considerable workload prevent him from keeping another, independent diary. He gathered material for a precise ekphrasis of the power structures in the Führer's headquarters. There would not be enough time for Hartlaub to develop the writings of his diary into a novel about a "war diarist."[106] In the last month of the war, his trail is lost in the confusion of the house-to-house fighting in Berlin. His private writings, on the other hand, have largely survived. Someone who has to keep the war diary of more than eighteen million Wehrmacht soldiers can only register his own dissolution into a mere "writing finger, reading eye, observing conduit."[107] Judged according to Edmund Husserl's concept of an inner sense of temporality, Hartlaub's existence in the second restricted zone of the headquarters knows only protentions and retentions, but no immediate presence filled by any intentions:

Time here, that's something in itself, it has nothing to do with ordinary time, more like with eternity. It is always the same day . . . and the same year, which represents all six war years, all moments of the war are amassed here, the past ones are not really past, and the present ones are not fully there, the calendar is only used for understanding with the external world, for the setting of x-days, report dates, but here, within the restricted zone, it is only valid to a limited extent.[108]

Even an explosion in the Führer's headquarters on July 20, 1944 does not tear Hartlaub away from his work routine. Only a poem written in retrospect and then discarded had the world "[stop] spinning for a single second,"[109] whereas the diary admits "upon hearing the explosion to have felt nothing of the sort."[110] An isolated explosion in the war can mean all sorts of things: a deer that has walked into a minefield, a dictator whose blood is pouring over general staff maps.[111] In war, meaning is long in coming—if it even arrives at all.

Führer Principle

It is the object of the historiography of war to provide the dispersed soldier-subject a leviathan-like general overview that cannot be gained from the narrowly private perspective. As little as the soldier-subject exists for himself, he grasps the ongoing events of war just as little. Rather, the elements of war are fully absorbed into retentions and protentions; thus the past battle is only and first reconstructed in order to be instructive for future ones. In an ongoing battle, the soldier trained in war games is therefore fighting battles long past, and indeed in the way these actually should have gone. At the same time, the currently ongoing battle will come to light only once it becomes the basis for planning the coming battle. Paul Virilio, who performed his military duty as a cartographer and called his service in the Algerian War his university,[112] has lucidly recognized the temporal dimension of the war game and declared the "dromoscopy" outlined by him as a "sophisticated form of *Kriegspiel* . . . in some ways a video game of speed, *Blitzkriegspiel*, where the military practices of the major state are continuously perfected."[113] Inherent to the dromoscopy is the vision of a world lost as soon as it is perceived.[114] With Virilio's theory of dromoscopy, it is possible to explain how the restriction of the Reichswehr to a hundred-thousand-man army, initially imposed by the victorious powers, could ultimately lead to a vast strategic potential. Armament and attack measures that have been virtually played through over years offer the adversary neither politically nor militarily concrete attack points or possibilities to take countermeasures. This forces the very development of the power of command, which is all that remains as a key to the creation of facts. On the one hand, the power of command is thereby exempted from the status of the virtual; on the other hand, it too can be virtualized insofar as command structures regulate how and when the power of command is passed from one subject to another and thus provides not only the dissemination of an order but also the power of command itself. Virtualized power of command is therefore commanded power of command, which must manage the complex of obeying orders at the same time as commanding.

Such command structures presuppose a particular Führer type. In Walter Elze's "military history section" at the Friedrich Wilhelm University, this too was worked on—indeed, even before the establishment of the National Socialist Führer state. With Friedrich von Cochenhausen, who enrolled in the historical seminar at the Berlin University in April 1932 for two semesters,[115] sessions of the military history section included a participant who had nothing more to learn in questions of military education. Before the

First World War, Cochenhausen had already been a teacher at artillery and engineering academies, was detailed during the war to the Great General Staff, and after 1920 was an advisor to the Ministry of the Reichswehr in the army training section (T4). With Cochenhausen, who was appointed general of the artillery and received a Ph.D.,[116] clear teacher-pupil relations disappear as much as the boundaries between military training goals and so-called academic educational ideals. Before Cochenhausen wrote a brief essay in Elze's historical division on the Austrian Chief of the General Staff, Conrad von Hoetzendorf,[117] he had encouraged young Reichswehr officers in the Reichswehr ministry to collaborate on the text "Führertum" ("Leadership") and had published it.[118] Directly after his brief engagement in the historical seminar, Cochenhausen published as "Führerschulung" ("leadership training") an instruction manual on the directing of map exercises and war games together with a suitable game apparatus.[119]

A career like that of Cochenhausen shows how much Michel Foucault erred when he presumed that "the reawakening of Frederick and of all the nation's other guides and Führers," took place in order to allow "State racism to function within an ideologico-mythical landscape,"[120] for the relevant military sources of the last days of Weimar reveal that the concept of the "Führer" is a military-technical term with a very particular function: in the concept of the "Führer," the difference between the claim to power, the exercise of power, and the participation in power is suspended in order to regulate this trinity depending on the situation.

Thus it is already in the beginnings of the Reichswehr that the designation "Führer" becomes a key concept. It therefore occurs within the time of the first German democracy. It stands for a specific form of organization that allowed Hitler himself to speak of the achievement-based "Führer principle"[121] of the Wehrmacht. After the Second World War, first the armed forces of the United States took up the doctrine of the achievement-based "Führer principle," then its business leaders.[122]

With the common tendency to equate the concept of the Führer with the Führer cult, a linguistic regulation measure taken by the propaganda minister of the "Third Reich" threatens to gain a validity retroactively that it had never possessed in such an unrestricted fashion. Anyone who merely inquires is disabused of that notion by as relevant a source as the *Handbook of the Modern Military Sciences* published by Cochenhausen in 1936, under the heading "Führertum": "Not only those on whom nature bestows abundant gifts are called to be Führer. Rather, everyone can be a usable Führer."[123] What is invoked here is not a charismatic figure of the Führer to which the mass submits, but one that the mass develops in its own ranks in order to exploit untapped potentials of self-organization.

In order to grasp the change in meaning and function that the designation "Führer" undergoes, it is worth looking at the military handbooks from Moltke the Elder onward into the 1930s. Soldiers are designated as Führer who assume a command contingent on the situation and then give it up again with the execution of their mission. Their authority is assigned to them from without. They are the beneficiaries of the state of exception. They can neither derive their warrant to exercise power solely from themselves nor secure it permanently. Moreover, during military training, the designation "Führer" helps overcome the precarious circumstance that, particularly in war games, positions are assumed for the purpose of practice that officer candidates and officers are not entitled to occupy according to their rank. Thus war game manuals sometimes contain the sentence "The war game is only as good as its Führer." With Cochenhausen, it consequently becomes apparent how an only moderately positively charged concept receives a new functional character and leads to collective internalization. If the role of the Führer was initially limited to a particular situation and the Führer was always aware of his replaceability, now an escalating momentum develops such that every state of exception—and this alone—brings with it the possibility of maintaining that state and thus the power of command once it has been attained.

Later, during the Second World War, it is the leadership qualities demonstrated in exceptional situations that are honored with medals and promotions regardless of length of service.[124] Because promotions do not follow a predetermined schema, each military rank receives its value above all through the possibility of reaching a still higher one. In other words, military officers are driven to leadership, to *Führung*, by states of exception.

Because Führers must emerge in exceptional situations directly from the circle of their followers, "differences of rank"[125] should not carry all too much weight. Rather, it must be ensured that officers of different ranks are equally well equipped for leadership tasks and situations. The common sphere of action is fragmented with mutual, if not equivalent, dependencies: "The greater the Führer's sphere of action, the less he is able to survey everything. He needs Unterführers, to whom he gives orders. Division of labor is essential."[126] The training of general staff officers, as so-called *Führergehilfen*, therefore demands the judgment of "the decisions of the Führers and Unterführers . . . according to the situation or the impression under which they acted. It is important to determine whether the decisions that diverged from the order were in the spirit of the higher Führer."[127] During their training, General Staff officers must understand this recursive function of Unterführer, Führer, higher Führer—at the end of which stood that Führer and Reich chancellor who attached less and less importance to bearing his second, constitutionally granted title at all.

With recourse to the doctrines of the leading "brand technicians" of his time,[128] Josef Goebbels might have created the designation "Führer" as a monopoly for the person of Hitler[129]—however, by no means from a "more or less naturally grown epithet."[130] The dimension of linguistic politics that opened up here is more significant. To order via the propaganda ministry that phrases like "Führer des Betriebs" (manager of a factory) are to be avoided, and at most words like "Betriebsführer" (factory-manager) will be tolerated, is one side of the coin. The fact that Goebbels's linguistic regulation nonetheless encountered limits is the other. Hitler was aware of this:

If in the present conceptual overlaps occurred, such as the captions under photographs: "next to the Führer, the Oberführer so-and-so, his adjutant" . . . that did not matter, as long as he lived. But once he was gone, that would have to be changed and the term "Führer" would have to be elevated to a unique concept. Ultimately, no one would think to call a streetcar driver [*Straßenbahnführer*] a streetcar Kaiser.[131]

That the "Führer" standing next to the "Oberführer," from a purely nominal perspective, does not look good for posterity was the fault of the National Socialists themselves with their linguistic regulation. Before Goebbels elevated "the term 'Führer' . . . to a unique concept"[132]—though the original idea went back to the Reichsarbeitsführer Hierl[133]—the armed organizations of the National Socialist Party used the designation "Führer" consistently in the designation of their ranks.[134] In contrast, the Reichswehr had, with few exceptions, spoken of Führers only in internal linguistic usage—even if, as mentioned, they did so extensively.[135] Even after the transformation of the Reichswehr into the Wehrmacht, there continued to be talk of Führers informally, while officially the nomenclature of the Prussian General Staff was revived.[136] It can be assumed that the National Socialist organizations did not merely take up an internal designation of the Reichswehr and elevate it to an official component of their rank names—rather, with it they took up above all the principle of rapid growth. Ultimately, the Reichswehr had only just demonstrated how Unterführers can be led to the tasks of Führers and Führers to those of Oberführers, and so on, in order to promptly multiply an army. That Goebbels succeeded in reserving the word "Führer" as a title for a single person is a legend; its durability may well represent Goebbels's actual propagandistic achievement.

The dilemma contained in the designation "Führer" finds expression in the announcement that the circle of conspirators surrounding Claus von Stauffenberg intended to radio to the military district commanders after a successful assassination of Hitler: "The Führer Adolf Hitler is dead."[137] This communiqué would be followed by a report giving the impression that Hitler had fallen victim to a power struggle within his own party. At that point, the plan against internal unrest received by all military district commands was to be implemented via the password "Valkyrie." Even if—apart from Hitler himself—only Colonel-General Friedrich Fromm was authorized to issue the password, Stauffenberg as his representative could make use of the password at least by telegraph. According to the plan, military and economic facilities, telephone and telegraph offices, radio stations, transport facilities, and so on were then to be secured by a reserve army.[138] In other words, the conspirators were relying on "the absolute obedience of their subordinates and comrades, and attempted to pursue their treasonous goals not *against* the military apparatus, but *through* it."[139] With the announcement of the death of the Führer, the conspirators, of all people, wanted to speak for the last time in the name of an indivisible power of command in order to reinstate the older Führer principle,[140] for it was this very principle that saw in the death of a Führer the emergence of the situation that demanded the compulsory and spontaneous appointment of a new Führer.

Only in the *Handbook of the Modern Military Sciences* of 1936 do the countervailing tendencies become evident that converge in the designation "Führer." In a contribution written by Carl Schmitt, the development—construed politically—amounts to the "National Socialist Führer state,"[141] and in Cochenhausen's military-operational analysis, it amounts to a Führer army. The fusion in the collective singular of state and army is based on a system of self-similarities and the division of labor, in which everyone—in practice, starting at a certain rank—must be able, in principle, to take the position of the other. That is exactly what Hitler formulated in his speech before the Reichstag at the beginning of the Second World War:

> I want to be nothing but the first soldier of the German Reich! . . . Should anything happen to me in this battle, then my first successor is party comrade Göring. Should anything happen to party comrade Göring, the next successor is party comrade Heß. You will then be just as duty-bound to blind loyalty and obedience to him as Führer as you are to me!
>
> In case anything should happen to party comrade Heß as well, I will then convene the senate by law, which will elect the most worthy, that is, the most brave, from its midst.[142]

The power structure is no longer based on a hierarchy of offices, in which power is distributed from the top down into a broader base with increasing limitation. Rather, it is defined by a circle of substitutions, which are triggered and propelled only by exceptional cases. It is no longer he who maintains his command in the exceptional case who becomes the state sovereign. Rather, one henceforth becomes sovereign in the first place only in the state of exception.

Chains of Command

Until Seeckt took over the army command shortly after the First World War, the unwritten law had always been that only officers with war experience maintained the capacity for defense and aggression. Periods of peace that lasted longer than twenty-five years put the General Staff on the alert, because then the longest-serving officers who had fought in the previous war retired from service. Seeckt, however, whom the Treaty of Versailles granted no more than four thousand officers in building the Reichswehr, discharged—against considerable resistance—primarily battle-tested, seasoned commanders and drew together a disproportionate number of officers from the planning staffs of the General Staff. An authority that invoked war experience and thus a state of exception could not be undermined by subordinates who did not possess such experience. Things were different in an army that relied consistently on training and map exercises. With the successful transmission of tactical and strategic knowledge, the differences in level between trainer and trainee diminish and ultimately—when advantages in knowledge and planning are their sole measure—disappear completely. And it is not only knowledge that is imparted to the subordinates, but also the knowledge of the technique of its transmission and attainment.

An army of trainers trained not only officers but also officer trainers. Therefore, he who rose in rank did not merely assume the position of his superior, but also officially took on a function that he had already previously had to enact virtually. In particular, Seeckt found in the war game a medium of military development for his Führer army that did not disavow authorities:

The officers [*Fuehrergehilfen* in the original] had to act as directors and leaders [*Fuehrer* in the original] in constant alternation. For every year of training at the War Academy, each commanded officer had to set up and direct at least one war game and a terrain briefing.[143]

While every single prospective officer had to take on the role of his superior, as game director as much as in his real command function, the troop formations as a whole were trained according to the same principle:

For the classification of the participants, the following shall serve as a norm: Every leader [*Fuehrer* in the original] leads in the game the same formation that he leads in actuality or the next higher one, so that, for example:

a young battalion commander leads a battalion
an older battalion commander a regiment
a young regiment commander a regiment
an older regiment commander a division etc.[144]

Not only the decisive officers test themselves in the function of their own rank and a higher one in the war game, but also the complete troop formation:

The framework of the game is purposefully chosen so that it is a level higher than the formation of the staff with whom it is being played, so that:

with a battalion it is a game in a regiment framework,
with a regiment a game in a division framework,
with a division a game in a corps framework.[145]

The military historian Martin van Creveld has pursued the question of why the Wehrmacht forces—in comparison with the American army and others in the mobilization of the same underlying amount of men and weapons—as a rule "were 20 to 30 percent more effective than the British and American forces facing them," even regardless of whether the battles as a whole were waged with a superior or inferior number of troops.[146] Part of his answer was that the mission-type tactics of the Wehrmacht left the method of the execution of a mission to the respective lower-ranking officers and endowed them with all the necessary power of command. Unforeseen situations or even the loss of a troop leader, van Creveld concludes, could be managed better in a unit that was characterized by "independent thinking"[147] than by an army like the American one, whose command technique relied on very precisely issued orders and scarcely surmountable hierarchies that had difficulty adapting to altered conditions. With the Reichswehr, the message of the psychoanalytic movement had been largely received:

The loss of the leader [*Führer* in the original] . . . brings on the outbreak of panic, though the danger remains the same; the mutual ties between the members of the group disappear, as a rule, at the same time as the tie with their leader.[148]

But although disciples of psychoanalysis recognized in war trauma after the First World War an immense field of treatment, military trainers— supported by the psychiatry of their time—had long ago proceeded to train a Führer type who was already prepared in the war game—that is, prophylactically—for traumatic scenarios.[149]

In other armies, this approach had been discovered only partially: Van Creveld mentions the case of an American officer and social scientist who came across the fact in the German manual for unit command "that officers of all levels . . . were forced to analyze their own situation as well as that of their next higher command level" and noted that "higher command level" probably stood, due to a mistake, for "lower command level."[150]

In wars that are waged over borders and across them, battle techniques may well arise least of all from strategic acumen, which is a matter for one and only one nation. The American officer and social scientist who did not, as assumed, come across Prussian blind obedience would have been perhaps still more surprised if he had learned that an order that required from the "youngest soldier upwards the total *independent* commitment of all physical and mental forces [emphasis added]"[151] probably first emerged among Hessian troops returning from the American war of independence.[152]

But it is not necessary to go back to the American war of independence to come across the principle that—contingent on the situation and in particular when there are losses of high command—ideally puts every unit in the position to take over the command. The U.S. Navy in the First World War relied—in contrast to the American ground forces of the Second—on this very principle. Warren McCulloch, who would receive attention as one of the initiators of cybernetics and as a neurophysiologist, learned of the principle as a young marine officer:

[When] America joined World War I, McCulloch, given a family history of patriotism, wanted to join the Navy. He therefore moved to Yale University, where he joined the Officers' Training Program. There he divided his time between officers' training courses and time on a ship, combining "marlin spike sailing" and signaling by semaphore. Perhaps some of his ideas about coding in the nervous system were shaped by his concern for coding messages and transmitting them from ship to ship. Another idea from the World War I Navy, to which we will return, was what he refers to as "redundancy of potential command." In a naval battle, there are many ships widely separated at sea, and normally command rests in the ship with the Admiral. But if some fighting breaks out or some crucial information becomes available locally, then temporarily the ship that has that information is the one with command. This notion of redundancy of potential command, rooted in McCulloch's experience in World War I, came in the 1960s to yield the view that the nervous system is not to be seen as a pure hierarchy but rather operates by cooperative computation.[153]

The principle of the "redundancy of potential command" opens up an epistemological space that no longer draws a strict distinction between inside and outside, human and machine. Cybernetic models in their systems of equations illuminate human and machine alike. The knowledge of the redundancy of potential command, as manifested in naval battles, is therefore embodied for McCulloch in reticular formations of the brain stem.[154]

Within the transference of knowledge, the concept of information takes on a key role. Information becomes the measure that maintains human as much as mechanical communication in "the presence of noise"[155] through redundant coding. What is signal, what is noise, and what is their conduit medium can remain open for the time being; what is significant is the differentiation of these variables in the first place. In Claude Shannon's communication theory, this insight is incorporated into a mathematical model. The most impressive object lesson for the underlying paradigm that will determine the theoretical standard of the postwar era is reported by Field Marshal Wilhelm List, though with the battlefield in view: "It remains . . . to consider that the path from the head of the Führer to the acting troop is far, that much time elapses from the point of the decision-making to its implementation in the deed. Frictions of various sorts, misunderstandings, mishearings, omissions, among other things, prolong this time."[156] For the hour of the birth of cybernetics, McCulloch held a meeting in the winter of 1943–1944, which brought together biologists, physicians, engineers, and mathematicians, such as Norbert Wiener and John von Neumann, to extrapolate from the communication methods of living organisms to technological solutions. A fictitious discovery crystallized as the object of the meeting: "two hypothetical black boxes," captured from the Germans.[157] The first box is opened and immediately explodes. The second, which—like the first—has "input" and "output" interfaces, leads to the illuminating question: "This is the enemy's machine. You have to find out what it does and how it does it. What shall we do?"[158] Norbert Wiener proposed feeding the input interface white noise: "You might call this a Rorschach." Neumann opposed that approach with "feature-filters."[159] Ultimately, they agreed to meet more often in interdisciplinary cooperation. The groundwork had been laid for the notorious Macy conferences and with it for cybernetics as a new leading science. Though today cybernetics may be history, the idea that neurons *fire* in our heads has entered not merely the history of science, but also common parlance.

5   Higher Mathematics and *Nomos* of the Earth

Higher Mathematics in the General War Academy

Conditions of war are clearly suited to orienting forms and processes in all areas of life toward a determinate *telos*—at least more so than is possible in peacetime. The concentration of human life on the attainment of fewer—but for that reason all the more sharply delineated—goals converges with mathematical disciplines. Furthermore, it allows other disciplines and more distantly related discourses to seek a connection to their rigorous calculations and methods.

This state of affairs can be inferred, in any case, from a letter that mentions two civilian actors who are nonetheless decisive for the arms industry: "Last week President Conant is reported to have said to President Jewett: 'The last was a war of chemistry but this one is a war of physics.' To which President Jewett replied: 'It may be a war of physics but the physicists say it is a war of mathematics.'"[1]

It is thus not the generals who here venture to assess the ongoing world war, but two scientists: James Conant, president of Harvard University, and Frank Jewett, president of Bell Telephone Laboratories and the National Academy of Sciences. Both were appointed by the science functionary Vannevar Bush as members of the National Defense Research Committee (NDRC). Under the aegis of these civilian researchers, the NDRC would get its biggest militarily decisive research initiatives off the ground: a microwave radar system and, in Los Alamos, the atomic bomb project.[2] Consequently, it must be noted that the Second World War was a war not only of military officers but also of civilians—and not only in the role of passive sufferers but also in that of activists. But the extent to which the Second World War became a war of mathematicians is more difficult to fathom.

Are not high-frequency technology and the atomic chain reaction the specialties of physicists, for whom mathematics—however complex and innovative it may be—is only an auxiliary science?

In order to pursue this question, it is again worthwhile to focus on cybernetics. Scholars rightly discern its beginnings during the Second World War: From a historical perspective, the ontology of life designed by cybernetics goes back to an ontology of the enemy.[3] However, the concept of positive and negative feedback—established by Norbert Wiener, the co-founder and namer of cybernetics—dissolves a dichotomously understood friend-enemy schema, for according to Wiener, nothing stood more pressingly on the agenda during the Second World War, shortly before the founding of the "Teleological Society"[4] club, than the construction of an "anti-aircraft predictor," which would consolidate the physics of flight (spaces of play that once would have simply been called "nature"), the individual aviatic possibilities of enemy pilots, and one's own anti-aircraft defense into a single feedback system. According to Wiener's mathematical model—the implementation of which foundered on its own complexity and the technological feasibility at the time—the three elements are reflected only as a time series of past measured data that enable an extrapolation to a future system state. Whether this predicted state is read as the launch position of an aircraft or as the danger zone depends solely on the standpoint in relation to a line of demarcation, which now proceeds between earth and sky.

A more functional anti-aircraft predictor, if it came into the enemy's hands, could not only be deployed as such, but also as an "anti-gunfire predictor." The recognition of this simple symmetry brings to light the fact that Wiener's system can be used against itself. Friend and enemy are, within the framework established by Wiener's theory, equally responsible for differentiating one and the same system. To put it more generally, mathematicians brought their knowledge into the Second World War in a fashion that did not simply lead to new weapons of greater destructive potential. It is not weapons as such that they forge, but systems and platforms, within which battles are fought out and which are not directly aimed at each other like weapons. While physicists and chemists literally participate in the material battles, a different, primary task is assigned to the mathematicians. They reveal fundamentally new battlefields by making spaces controllable that otherwise withdraw from visibility and already inherently tend to exclude life in a lethal fashion: spaces of atomic radiation; darkened and sealed-off spaces, such as those that are surrounded by bodies of water, inhospitable land masses, or airspaces; and finally spaces that offer no orientation due to their sheer size. These spaces are first made passable, navigable, and communicable through media that are themselves withdrawn from the register of immediate visibility: sonar, ultra-short wave radio or centimeter-wave radar. Mathematically well-defined structures are molded to them. The technique that wrests from these media their spatial and temporal structural properties is none other than mathematics (and not—as a certain media theoretician from Canada who was well versed in literature but clueless about technology would have one believe—only some other medium). Mathematics is not a medium. It must be taught and learned, and it must be communicable and, above all, demonstrable. Mathematics may be distinguished by its abstractness, but it nonetheless requires forms of evidence and visibility.

If one searches for a moment in time when mathematics unfolded in all its abstractness regardless of its applicability and for this purpose occupied a specific space proper to it, then—at least focusing on the German context—this is rather simple to determine. The year and the hour are documented in which mathematics in German regions proved for one last time to be incommunicable: Franz Neumann, who would gain prominence as the founder of mathematical physics in Königsberg, had left for Berlin in the winter semester of 1817–1818. He hoped that, unlike in Jena, mathematics would be taught there. Neumann records in his lecture notes:

The professor entered the auditorium, stood at the lectern and wrote, with his back to us, mathematical formulas on the board without interruption, spoke not a word, continued to write until the time was up; then he bowed to us and left. On the second day only three listeners came. The professor again stood at the board, again wrote mathematical formulas on it without interruption, spoke not a word, took his bow, and the second lecture was over. The third day only one listener came besides me. The professor appeared, went to the lectern, turned to us and said: "You see, gentlemen, the lecture course has fallen through," took his bow and left.[5]

Tres faciunt collegium was already the rule at the medieval university. After three days, the professor counts the remaining two attendees and cancels the lecture with the only arithmetical act destined for comprehension. Thus, one of the few mathematical lecture courses in German regions—if not the only one—ended before it had really begun.

In retrospect, the history of science requires no more than two sentences in order to determine the beginning of the institutionalization of mathematics in Prussia, which would set the tone for all German states: "The heyday of mathematics began with the appointment, carried through by A. v. Humboldt against the will of the [Berlin] faculty, of Dirichlet, who in 1828 first came to the War Academy. . . . Dirichlet created the type of mathematical lectures still common today."[6]

Less than a generation after Gustav Lejeune Dirichlet, the Berlin University is not merely one of the few possible addresses, but the prime address for mathematical instruction, which is ensured above all by the triumvirate of Eduard Kummer, Karl Weierstraß, and Leopold Kronecker. Kummer and Kronecker are pupils of Dirichlet. He himself had still had to leave his hometown, Aachen, which belonged to the Prussian Rhine Province, in order to be able to study mathematics. Higher mathematics was taught to him only in Paris. The fact that Dirichlet's unprecedented career as a mathematics instructor nonetheless began at the General War Academy in Berlin cannot be explained with a biographical gloss. Above all, it is in Dirichlet's letters that the connection between General Staff education and the mathematical seminar comes to the fore. This connection became a deep-seated sediment that formed the basis of the military-technological complex of the twentieth century. Ultimately, mathematics first found the form of its knowledge transfer with Dirichlet's career at the General War Academy in Berlin and went on from there to conquer the university.

Likewise, it is no accident that Chief of the General Staff Karl von Müffling obtained an audience with the Prussian king in order to point out the unfortunate situation in the impartation of mathematics: "I told the king that the state instruction in mathematics in other nations begins where it concludes here, that though we will always find mathematicians among us, the people as a rule become so brusque and one-sided, as a result of the fact that they have to educate themselves through private study, that the state then ultimately has no use for them."[7] The fact that mathematicians are to be regarded as a rare species whose rarity often makes them strange and of no use for the state was something Müffling intended to change. In Alexander von Humboldt he found his closest ally. Though Humboldt's brother Wilhelm, as chief of culture and education in the ministry of the interior, had put the Berlin University on a new, humanistic basis, he left untouched the medieval division of the departments into theological, medical, philosophical, and juristic. That is why, in Müffling's criticism of the university and academy after his plan to found an *École polytechnique* in Berlin met resistance, the atmosphere no longer sounds so neohumanist at all. He found "that our German philologists are just as intolerant as the Jesuits and that there is an alliance not to allow mathematics to develop."[8]

Thus there was nothing to be done with German philologists. Nonetheless, Müffling and Humboldt not only envisaged the founding of an *École polytechnique* along Parisian lines, but they also tried to woo Carl Friedrich Gauß, Germany's mathematician-prince, away from the Kingdom of Hanover. Decades later, a letter from Humboldt to Gauß indicates that they wished to engage him not only for the Berlin University or the Prussian Academy, but also geopolitically for "his fatherland."[9]

For a moment, Müffling and Humboldt's plan to appoint Gauß even seemed to be working. Müffling—authorized by King Friedrich Wilhelm III—saw to Gauß's demands and promised to keep the appointment process strictly secret.[10] And Alexander von Humboldt used the 7th Congress of German Naturalists and Physicians in Berlin as a pretext to invite Gauß to Berlin beforehand. They breakfasted at Humboldt's home. Along with Dirichlet, he requested the presence of such prominent researchers as Charles Babbage. The inventor of "the machine that calculates and prints,"[11] Humboldt let Gauß know, "is overjoyed at your arrival."[12] Gauß, actually known for his soberness, raved in his letter of thanks to Humboldt about the happiest days of his life. Then Müffling committed a tactical error. Though he granted Gauß's demand to be freed from teaching at the university, he miscalculated in his belief that Gauß would take pleasure in having a "great influence on the whole mathematical educational system of the state" and the still-to-be-founded polytechnic institute.[13] Gauß saw merely "ministerial affairs" ahead of him, which contradicted his only emphatic demand for as much space as possible for his own work.[14] While Prussia stumbled into a negotiation crisis, the representative of the English royal house in Hanover reacted promptly and bound Gauß further to Göttingen through the promise to support him more strongly.

Though the founding of the polytechnic school would ultimately fail, Alexander von Humboldt had made provisions. The breakfast with Dirichlet, Gauß, and Babbage was followed a week later by another. This time, Chief of the General Staff Müffling, Major Joseph Maria von Radowitz, and

again court councilor Gauß and Dirichlet were invited. Major von Radowitz was already distinguished by the fact that—as Humboldt put it—he was
"as a Westphalian a graduate of the *École polytechnique*."[15] For the founding
of a polytechnic school in Berlin, he therefore seemed indispensable. Consequently, after all the efforts to found a school had come to nothing, he
took over as the director of studies at the General War Academy, which
came closest to achieving what they had hoped for from an *École polytechnique* in Berlin. After two world wars, nothing remains of the General War
Academy, and most of its records and documents ultimately went up in
flames when the military archive suffered a direct bomb hit in April of
1945.[16] But the nearly total disappearance of a college that had emerged
in part at General Scharnhorst's instigation should not belie the fact that
it was initially placed on the same level as the Berlin University. The
General War Academy and the university opened their doors on the same
day in 1810.[17] From the outset, the educational offensive that came about
with Baron Karl von Stein zu Altenstein's reforms was a double one.

The question remains of how Dirichlet first came to be among influential Prussian military officers and court councilor Gauß at Humboldt's
breakfast table and ultimately attained a position at the General War
Academy. When Humboldt invited the twenty-three-year-old Dirichlet to
breakfast, he instructed him that he should just happen to carry under his
arm Major von Radowitz's "Handbook for the Application of Pure Mathematics."[18] Unfortunately, no copy of the book—the title of which today
seems to be an oxymoron—can be found. But according to the sources at
the time, it consisted of tables of trigonometric formulas that, once collected, awaited their application at the General War Academy. Properly
deployed, trigonometric skills could be used to advance land surveying,
which made possible the production of general staff maps. They could also
be employed for the calculation of standard values in the artillery.

But Dirichlet could not have been more remote from a mathematics
that sought its application in an open field. For his whole life, his mathematical operations would not go beyond the space of his desk or the
blackboard. He had already received attention in Paris with a partial proof
of Fermat's Last Theorem through the application of Gaußian methods.

Nonetheless, it should not be seen as a contradiction to regard Dirichlet
as a co-founder of mathematical physics. Rather, it is an indication that
mathematics freed itself from being pressed into the service of

experimental physics. Dirichlet's solutions to problems are based on boundary conditions that are rooted far deeper in hypotheses of analysis than in *physis*. On the whole, Gauß's number-theory book on Dirichlet's desk remained his most inexhaustible object of study. He first made it accessible to other mathematicians.[19] When Jacob Jacobi had to report to the culture ministry about the merits of his colleague and friend Dirichlet, he gave paramount importance to his applications of Fourier series to the theory of prime numbers.[20] Fourier series were capable of describing the spreading of warmth in a conductor—at a price, however, that dissolved a physical conception into nothing but discrete, graphemically very clear but nonetheless inconceivable curves.[21] And Dirichlet now applied the Fourier analysis, which had already removed physical phenomena like the spreading of warmth and sound from the sensory field, to the purely number theoretical determination of prime numbers. He began, in other words, to apply different mathematical branches to each other. Descartes had already famously subjected geometric questions to analysis. Following the example of Descartes, Dirichlet is praised for having now performed the "application of analysis to number theory."[22] But whereas Descartes's analytic geometry sought to capture the real methodologically, the systematic closure and inner differentiation of mathematics began to take place in Dirichlet's time—and with him as the driving force.

Against this background, the question arises all the more as to just what mathematical skills Dirichlet was supposed to impart at the General War Academy. A letter from Humboldt to Dirichlet outlines the problem:

Herr von Radowitz knows as I do that you have hitherto occupied yourself only a very little bit with the branch of applied mathematics that is connected to geodesy and the artillery. But modern physics, ballistics itself, amounts to analysis, and with your acumen you will soon understand more about it than Puissant and Poisson, who was recently forced to discuss the form of wheels and freight wagons. You will above all discern what is essential in this problem of projectiles, and what the positive data are that the experiment must provide. One does not ask that you yourself carry out or direct these experiments, one only wants you to indicate where you consider them necessary for the analytic calculus. Now, Herr von Radowitz cannot be useful to you and cannot hope to see to your being selected by the minister of war as long as he cannot show a small work by you in the field of analysis applied to ballistics. Without this work the minister and Prince August of Prussia, the inspector general of the artillery, will hold against your appointment that, though you are a significant mathematician, nothing demonstrates your wish to descend into their field of action. One toils here with certain analytical calculations of the rotation of hollow spheres, their deviation caused by air resistance. . . . Herr von Radowitz says that he himself has worked on that unsuccessfully, because he believes that he has not found the true analytic method; he therefore would like to be able to show the ministry a small sample of your work in applications of analysis to the movement of projectiles, present certain questions to you (through my mediation), through which you would explain your ideas to him as to whether perhaps due to the lack of experiments or rather of numerical data provided by these experiments an ultimate solution may be impossible.[23]

Humboldt feels compelled to remind Dirichlet in another letter of the necessity of a ballistic work and adds by way of explanation: "The problems of geodesy do not interest the minister of war and Herr von Radowitz. The trigonometric field belongs exclusively to the General Staff and General Müffling, who believes he is completely equal to the task."[24] And Humboldt urges Dirichlet a third time to deliver an analytic treatment of the ballistics of hollow spheres—that is, the precursors to grenades—because he now presumes that it has already been written. But nothing further has been passed down about such a work. Dirichlet's successor, Eduard Kummer, would take on the problem almost half a century later, shortly after the Franco-German war. However, he comes quite quickly to the conclusion that it "cannot yet be managed by purely mathematical means" and it is necessary to "turn to experiments," which he then did.[25] In Dirichlet's case, one searches in vain for works that proceed from physical experiments and not always already from mathematical constructs.

Nonetheless, Dirichlet receives an appointment to the War Academy. After the catastrophe of Jena and Auerstedt in October 1806, the effort was made there to teach mathematical and military operations with the help of one and the same technique. And that technique was called the applicatory method. The Napoleonic educational system already aimed for a *"heureux melange des études théorétiques avec les applications pratiques."*[26] Lectures were followed by a comprehensive course that taught pupils "mathematical drawing" and ultimately enabled them to prepare independently the "survey of terrain, buildings and machines."[27] The *École polytechnique* was followed by specialization at one of the *Écoles d'application*, the architecture and war academies.[28] It shows Humboldt's strategic skill that he directed the Prussian culture minister's attention to Dirichlet from Paris, by writing that the latter "would certainly prefer a position at a great [Berlin] *Gymnasium* to any appointment here at French war academies (for he can lecture in French as well as German)."[29] Thus, upon Dirichlet's appointment in Berlin, the culture ministry could believe that it had snatched him from the archenemy. Dirichlet was first assigned the modest role of "tutor . . . for application," practicing with officer candidates what professors of mathematics presented in lectures.[30] The mathematician Erich Lampe, who taught at the War Academy as the successor to Dirichlet and Kummer, did not forget to point out in his obituary for Dirichlet that "the operation of the seminar that the universities first introduced in general in the second half of the 19th century was immediately prescribed systematically during the organization of the General War Academy."[31]

Kleist's old friend Otto August Rühle von Lilienstern, as head of the War Academy, had charged none other than Radowitz with revamping mathematical education. Radowitz drew on the French "tutorial system," which he knew from his time at the *École polytechnique*[32] and called for the "dogmatic . . . lecture" to be accompanied by a "regulated self-occupation."[33] Rühle von Lilienstern could now emphatically recommend Dirichlet, for the latter was "educated in the polytechnic school in Paris and seems to be particularly suited to lecture in mathematics at the Royal General War Academy, because he is acquainted most closely with the *applications* method that has been carried out in the institution for two years."[34] Lilienstern's claim that Dirichlet had already become acquainted with the applicatory method in Paris reveals wishful thinking, for Dirichlet never set foot in the polytechnic school.[35] Dirichlet encountered Paris's great mathematicians, such as Fourier, Poisson, or Laplace, at the Parisian Academy of Sciences and at the Collège de France, which then—as now—granted anyone admission. Thus, upon closer inspection, Dirichlet scarcely served as a valuable source for an applicatory method along French lines. Whatever was introduced with the applicatory method at the War Academy in 1826 was a greater experiment than was acknowledged in Prussia. It soon turned out that, in the applicatory lessons, no tutor could or would practice even remotely the material of the lectures.[36] Before long, Radowitz criticized the fact that Dirichlet's pupils in the applicatory lessons supposedly expect "more constructions and not a merely analytic process."[37] With even greater horror, he realized that Dirichlet introduced infinitesimal calculus to the officers entrusted to him in the applicatory lessons at their request—material that belonged to higher mathematics and was therefore banished from the curricula of lectures at the War Academy.[38] It was feared that experienced officers who would prove to be unbeatable on the battlefield could fail an exam due to higher mathematics problems. In order at least to close the gap between lecture and applicatory instruction, the administration instructed Dirichlet to give both courses. He at once ensured that infinitesimal calculus was taken up in the curriculum.[39] Dirichlet, who was the same age as the officers he taught, represented a system of theory and praxis that in France presupposed the institutional interplay of two schools: the *École polytechnique* and the subsequent *École d'application*.

The "tutorial system," which in France was designed to practice analysis in application to semiotic practices, was changed by Dirichlet into an applicatory method that taught how to deal with those branches of higher mathematics that had not been initially treated in the lectures. From the applicatory instruction, a seminar ultimately emerged that did not merely repeat mathematical skills, but formulated a knowledge whose applicability was still debatable.

The Fabrication of Physico-Mathematical Objects

No sooner had Dirichlet finally been appointed to the Berlin University than he established a mathematical seminar there that was devoted to unsolved mathematical problems. In addition, he used lectures to present the latest research results, which was a novelty.[40] Even if he was prohibited for two decades as a professor extraordinarius from conferring doctorates on students, that did not prevent him from disseminating his teaching practices. His friend and colleague Jacob Jacobi took them up and founded in Königsberg the first physico-mathematical seminar. From the seminar a school would emerge, and from this school ultimately emerged the mathematician David Hilbert, who was called upon in the beginning of the twentieth century to give mathematics its own program. Dirichlet's successor Eduard Kummer ensured that the mathematical seminar became a firmly entrenched institution at the Berlin University. Still-unsolved mathematical problems, which could therefore not be part of the lecture, were now submitted to professors and students together in the seminar. And from the second half of the nineteenth century onward, more than almost any other place of mathematical activity, this very mathematical seminar gained a reputation that transcended national borders.

The transfer of the educational method from the General War Academy to the university ultimately turned the university from a place of knowledge reception to a site of genuine knowledge production in which theory and praxis coincide. All that remains for the four classical departments with their discourses is to revolve around a praxis that, at least in German regions, had long been kept external to them. Ultimately, the university is not a church, a courtroom, or a hospital. And as for philosophy, Felix Klein, for one, regarded mathematics as superior to it, because it not only pursues thought but also traces it back to its axioms and helps it attain application.[41]

For Berlin's educational system, Alexander von Humboldt had called in his time for "the first observatory, the first chemistry institution, the first botanical garden . . . [and] the first school for transcendental mathematics."[42] First and foremost, he got the last of these. As chemistry, physics, and biology laboratories and hospitals first began to move into the universities, the mathematical seminar had long since transitioned into fabricating "physico-mathematical objects."[43] With mathematics at the university, knowledge caught up to its object.

Turns of Mathematics

After Dirichlet, there are three paths for the application of mathematics. First, even abstract number theoretical concepts experience their application to geodesic and geomagnetic fields. It seems that the tellurian exploration, which begins to equal the rich tradition of astronomy, develops into a German specialty. Riemann surfaces and Hilbert spaces emerged from Gauß's differential geometric land surveys.

Second, mathematics does not only find its way to applications that penetrate spaces based on its semiotic operations and measurements and supplant purely phenomenal descriptions. From now on, mathematics creates its own objects, which it admits into a space to be mastered mathematically. Examples of this are instruments such as Gauß's heliotrope and magnetometer. These instruments, initially developed for land surveying, turn out almost incidentally to be an alternative to optical telegraphs and to lay the groundwork for new communication systems.[44] Francis Bacon's image of the experiment that interrogates Nature and forces her to speak still stems from a hidden metaphysical language. The consolation that Alexander von Humboldt in his old age will give in his last letter to the dying Gauß must be evaluated in an entirely different fashion: Gauß has more than anyone else "first given dependability, measure and wings to the electric speech now spreading over sea and land."[45] Humboldt's poetological formula encapsulates Gauß's instrumental capture of geomagnetism—which had made obsolete, formerly to Humboldt's chagrin, his own merely descriptive research—with instruments that henceforth imprinted their signals on spaces.

Third, mathematics is endowed with an application that pertains to itself. Only at the moment when the impartation of mathematics is itself regulated by clearly delineated procedures is a decoupling possible that differentiates between applied and pure mathematics. From this perspective, pure mathematics exists only from the point when it receives space in seminars and the posing of problems and tests demand its exercise as such. Only the type of practices, applications, and spaces differentiates between applied and pure mathematics.

What rapidly ceases after 1806—that is, after Prussia's military collapse—is the attempt to link military operations with mathematical ones in an idealizing fashion. Officers can only shake their heads when they recall the penchant of their instructors for mathematics:

"[One of the instructors had still based] the best number of the platoons into which a battalion should be divided [on the fact] that in a square front and flank had to be the same size and that he applied a quadratic equation, found x^2, extracted the square root and found a number 13.2415987."[46] In order to escape such nonsense, mathematics was elevated quite generally to a "science of intellectual education."[47] The battlefield of the eighteenth century still corresponded to the ideal of Euclidean space. Its curve-free surfaces offered a platform for the rule system of warring units. What was nonetheless humane about this form of battle and slaughter was that the applied mathematics found a limit when it took the measure of human beings confined to time and space, even when they went to the extreme. At the moment when abstract and advanced mathematics plunges into the tellurian, into land surveying and geomagnetism, the tide turns. From that point on, mathematics models a battlefield to be mastered in such a fashion that the figure of the enemy appears only indirectly, but as absolute.

In the establishment of spaces that permit the pursuit of semiotic operations in all exclusivity, the Prussian military and the mathematical seminar meet. And this very connection endures even when pure mathematics is not obviously enlisted in producing technologies. Dirichlet's mathematical seminar and Jacob Jacobi's Königsberg offshoot, the physico-mathematical seminar, create numerous fictive objects, such as Kummer surfaces and later Möbius strips or Klein bottles, which certainly seem as beautiful as they do harmless. Though the physical realizability of these objects is a condition, it is not crucial. What is decisive is that they emerge completely from mathematical operations, which are prescribed to them through closed expressions of analysis. But because mathematical investigations are, in advance, already less and less subject to an applicability external to mathematics, nothing but their ubiquitous deployment increases the measure of contingency that produces unsuspected circumstances. That is why, in 1874—that is, at a time when mathematics is still regarded as an auxiliary science—the mathematician Paul Du Bois-Reymond can already reply to the question "What is mathematics?" with the simple answer: "What is not mathematics?"[48] He represents—shortly after the Franco-German war—the assessment that also holds for the French mathematician Jean Dieudonné after the First World War; though the forms of pure mathematics are not directed toward any purpose, they already fill arsenals with symbolic directives in advance. And no one—not even a mathematician—can know how and whether they will find their way into the real.[49] Thus behind all the purposelessness of mathematics stands a pure utilitarianism that does not chafe against the deficiencies of the present, but rather always already wagers on something still incalculably to come. In contrast, Adorno and Horkheimer's critique proves to be downright naïve; apparently under the spell of the human catastrophe of the Second World War, they are able to see in the radical expansion of mathematics only the primacy of calculation. The fact that mathematics does not mean calculation, but rather making-calculable, escaped them. Making-calculable, however, presupposes the incalculable; one can even say with Alan Turing "the incalculable of calculation." Industrial wars and world wars taught the lesson that what matters is the incalculable entrance of the calculable. Leading mathematicians and military officers therefore have at least one thing in common: they conduct operations even when they have no idea where these will lead.

All the more fundamental for the securing of such a procedure are the traces that the operations leave behind. In mathematics, the intertwining of the operation with its recording has been cultivated more than in any other discipline. Every operation is almost totally subsumed by a trace—or, to put it differently, every operation nearly coincides with its graphemically elaborated trace. That the most recent contribution of mathematics always transmits a historical forgetting is due to the fact that mathematics constantly overwrites its history and concentrates on the moment of the operation. What Dirichlet, by order of Major von Radowitz, taught his officers by applicatory method was not, as in France, the application of analysis to construction and terrain drawings, but the reading of the graphemic trace of analysis itself. This turn toward itself, which was executed more radically in mathematics than anywhere else, was elevated into philosophy by Ludwig Wittgenstein, and therefore the final chapter is devoted to him. His dictum "The meaning of a word is its use in language"[50] permits no shift to a metalanguage that provides a theoretical overview. What follows is therefore the reconstruction of a game with mathematical language, in which nothing less than the life of reconnaissance officer Wittgenstein was at stake.

6 From Formula Games to the Universal Machine

The Double-Entry Bookkeeping of Reconnaissance Officer Wittgenstein

On April 6, 1916, the only war diary entry reads: "Life is a . . ." Under the date of the subsequent day, it goes on: "torture, from which one is only temporarily relieved so that one remains receptive for further agonies."[1] What one cannot speak about, one must pass over in silence. "An exhausting march, a night of coughing, a society of drunks, a society of mean and stupid people."[2] The gunner Ludwig Wittgenstein, declared "completely unfit"[3] for service by the Austro-Hungarian Army, left behind his privileged life in the familial circle of friends, in which representatives of Viennese high finance met as well as artists. He left behind Cambridge's elite mathematicians and finally his Norwegian hut, the refuge of his philosophical and homoerotic existence. Yet he never quite arrived in the war. First Wittgenstein stands at the searchlight of a captured Russian patrol boat and helps secure the Vistula on the Russian border. A first lieutenant who happens to hear of his mathematical and technical engineering education entrusts him initially with organizational tasks, then with the acting supervision of the artillery workshop of the Krakow fortress. For Wittgenstein that means "office work,"[4] often extending into the night. But the implementation of the manifested expertise founders on the lacking power of command. Wittgenstein's men refuse time and again to follow his orders. His superior then puts him in the uniform of a Landsturm engineer[5] until the war ministry intervenes, classifies the request for promotion as a presumption, and rejects it. After a transfer to the balloon section is also turned down,[6] Wittgenstein requests to be sent to the front with the 4th battery of the 5th Field Howitzer Regiment. The men, however, "with few exceptions," still hate him, the "volunteer."[7] Someone who, though he is exempt from service, nonetheless chooses war and not merely a military officer career—and who now gives orders, without having been ordered to serve in the war—appears suspect to his comrades. Wittgenstein has himself transferred again: "Tomorrow perhaps, at my request, I'll get out to the scouts. Only then will the war begin for me. And maybe – life too! Perhaps the nearness of death will bring me the light of life."[8] Only now has Wittgenstein arrived in the war, at the front line. The constant complaining about his comrades, who make him think of demands for duels,[9] now comes to an end in his diary. Instead, from that point on, only quick prayers are to be found. He is at a forward "observation post."[10]

The consolidation of light field artillery and heavy howitzer battery under a central command, tactically integrated balloon sections and forward reconnaissance officers are all inventions of the First World War. In its initial stages, its artillery still resembled that of Napoleon, before it ultimately found its way to forms and standards that still prevail in Western armies today.[11] The European powers used the years of peace before the First World War to increase the quantity of their ordnance and its penetrating power. The tactics of the artillery nonetheless remained at the level of the previous battles still in memory. No one could imagine that the infantry would advance on the battlefield without, together with its own artillery, having its eye on the enemy. But when the massive firepower was revealed in the first battles, all that remained was the withdrawal of the batteries from the adversaries' field of vision. The increased penetrating power of the shells promptly underwent a reevaluation: what now counted was the distance that the projectiles overcame. At 9,000 meters, it turned out to be considerably farther than ballistics experts had foreseen in their tables.[12] War now proves not to be the father of all things but the bastion of unforeseen and unrealized facts.

Of all places, Wittgenstein's first operational area as a reconnaissance officer, north of the Carpathians, becomes a center of this bastion—that is, a tactical field of experimentation.[13] What Wittgenstein's war diary records from that point on can be read as the "double-entry bookkeeping . . ."[14] of two experiments; one is the foray into a logic that "takes care of itself"[15] and thereby breaks with Russell's type theory, which must dodge onto a meta-mathematical level for its foundation, and the other is the record at a forward post, as it was only just established.[16] Both experiments culminate in a self-abandonment that seeks on the one hand to escape compulsive suicidal thoughts in the writing of rules for one's own life[17] and requires on the other hand meticulous compliance with tactical shooting rules so as to protect through covering fire above all one's own warring units and ultimately also one's own existence. Rules for reconnaissance officers do not, of course, assert a claim to eternity such as Alfred North Whitehead and Bertrand Russell already evoke in the title of their *Principia Mathematica*. Yet it seems worth the effort to reconstruct what Wittgenstein was toiling over during his two and a half years as a reconnaissance officer on the front and during several continued education endeavors, if not on Russell's work—with whom he incidentally ceased all communication at this time.[18]

Since the Brusilov Offensive had decimated more than half of the Austro-Hungarian Army in the Bukovina in the summer of 1916, what remained of it—a remnant to which Wittgenstein belonged[19]—stood effectively under Prussian command.[20] The Prussian supreme army command was itself advised in artillery tactics by Colonel Georg Heinrich Bruchmüller. His noteworthy second military career began on the eastern front and ended, honored with the Pour le Mérite, on the western front after Erich Ludendorff's Spring Offensive. Perhaps Bruchmüller heard Eduard Kummer lecture on the calculations of shell trajectories in his last years at the Friedrich Wilhelm University. It is certain, in any case, that he gave up his studies of mathematics and physics quite quickly and chose instead to join a foot artillery company.[21] His active period fell in post- and prewar years without his career ever developing on the fast track.[22] After a fall from a horse and a subsequent nervous breakdown, Bruchmüller retired from active duty. Only when the front ranks were thinning out in the First World War was he conscripted again; he took over a command far below his rank and distant from the front. Bruchmüller's moment had come in Przasnysz in the summer of 1915, when he tested for the first time an artillery fire that pushed toward the enemy positions in several phases and enabled one's own infantry to move up "behind a mask of smoke and dust."[23] The notorious creeping barrage, as it was called, would revolutionize positional warfare on both fronts. The "most precise preparation of infantry and artillery for this combat operation, as became the rule in the east from the beginning of 1916 and in the west probably from the beginning of 1918" deserves particular attention here too.[24]

At Lake Naroch in White Russia in April of 1916, the creeping barrage achieved a breakthrough against a numerically superior Russian Army, and in Galician Tarnopol in 1917 it achieved, no less effectively, a counterattack. Ultimately, Bruchmüller's central command in Riga—which coordinated the creeping barrage together with several divisions and various artillery and shell types—played no small part in the shattering of the tsar's army. In Riga, Bruchmüller's reputation preceded him and, for reasons of secrecy, required an elaborate journey from Galicia via Berlin to the fortress city.[25] His name was simply regarded as a synonym for breakthrough intentions: The army knew him only as "Durchbruchmüller" (*Durchbruch* is German for "breakthrough")—a name that Ludendorff had carved in stone and that would stand on the wooden cross marking his grave in 1948. Still, Bruchmüller does not embody the esprit de corps of the elite. The history of the military elite in the First World War finds its high point with the development of the assault battalions (*Sturmbataillone*).[26] Nor, however, were assault battalions invented by the General Staff. They formed spontaneously in view of positional warfare.[27] The assault troops (*Sturmtruppen*) could be sure of their top position in the military hierarchy because their initiator Willy Rohr had created them from such a position.[28] If Bruchmüller brought together with his own artillery tactics in Riga what had emerged on the western front in the infantry tactics of the assault troops, it is not the supreme command of the military elite that should be discerned in that fact but the "unique result of an autopoiesis"[29] of the battlefields. Bruchmüller had given up his command in 1915 and was from that point on, strictly speaking, merely a retired colonel who temporarily occupied the position of an artillery adviser—even if crown prince Friedrich Wilhelm summarily instructed his army group during the Champagne-Marne offensive to equate Bruchmüller's recommendations with the orders of the supreme army command.[30] And Bruchmüller inculcated in the artillerymen a perspective that valued the gratitude of the infantry more highly than medals and honors, "which only individuals can receive for the totality."[31]

Perhaps only an adviser without high rank and command could propose detaching the various batteries from the autonomous responsibility of their commanders and coordinating them centrally from above divisions.[32] But here too all commanders could do was give orders that issued from weeks of meticulously developed attack plans and were tailored only to the specific positions and weather conditions. More easily than commanders, however, plans can include the suspension of a centralized power of command when centrally coordinated actions—such as the artillery fire brought together into the creeping barrage—decompose into mere individual actions, for example, in order to secure conquered positions. Likewise, Bruchmüller's plans included an interplay of command leadership between artillery and infantry, insofar as a specific situation demanded it. The greatest advantage of a form of warfare that consistently created facts initially on paper and not first on the battlefield clearly lies in its element of surprise. Instead of shooting and calibrating for hours or even days at known targets with shell trajectories that, due to prevailing influences, deviated to a greater or lesser extent from their standard values, and thereby possibly betraying one's own positions and intentions to the enemy, Bruchmüller advocated the method of his captain Pulkowski, who suggested systematically investigating in advance the most diverse influence variables. From the beginning of the battle onward, the artillery commanders now shot not only at enemy positions that were hidden from view, but also by means of a method that instead of their empirical knowledge had recourse to a scientific system.[33]

Because everything nonetheless depended on the execution of the plans, Bruchmüller introduced extensive briefings before every offensive in the infantry and artillery troops. Feedback thereby took place not only between soldiers in the ranks, who otherwise had to submit to a unidirectional command hierarchy, but was also intended for the battlefield, as soon as the infantry signaled to the artillery with flares that the creeping barrage should transition into the next phase. Even the horror experienced by the infantry would be applied through feedback to enemy positions as a psychological warfare measure. Efficiency no longer meant attaining a total physical annihilation through weeks of artillery fire, as had become common, and ultimately capturing only minimal strips of terrain. Rather, it now meant the neutralization of the enemy—its physical and psychological paralysis, while one's own material and troop forces remained spared as much as possible. Bruchmüller preached the shock of the first wave of attack, instead of relying on the eventually stoically accepted barrage fire. In rapid alternation, he had enemy positions shelled, but with an arbitrariness that scarcely allowed the enemy infantry to come out of cover even during breaks in the firing. The contingency of a possible death was thereby given—along with the spatial scattering of the shells—a temporal dimension. In Riga, Bruchmüller ultimately used gas instead of explosive shells, which were growing scarce. The gas, only slightly heavier than air, even infiltrated underground positions and was mixed in such a way that tear gas penetrated behind the Russian gas masks and made them unwearable. Additional lethal gasses could then do their work unhindered. The use of gas led to more wounded and less dead, which was regarded as a tactical advantage, because the high number of victims that needed medical attention not only meant a loss of combat power, but also bound additional forces that were necessary to manage logistical problems, to say nothing of the moral dilemmas that they caused the enemy. The efficiency of neutralization as opposed to simple destruction consequently amounted to the delay of death—only, ultimately, to reign over it all the more powerfully. In the end, efficiency did not even stop at one's own men: the gratitude paid to them did not come without the condition of putting their lives at stake in a calculated fashion, for the artillery could not simultaneously shell all of the enemy positions that the charging infantry battled, which is designated in the technical lingo as a "target-rich environment."[34] On the basis of "plans produced at a particularly large scale, in which the individual time periods of the target combat are represented," infantrymen therefore had to internalize a course of battle that had them charge enemy positions even when support by their own artillery was in the plan only at a later point in time.[35] But Bruchmüller's tactics did not only demand of the infantry that they entrust their lives to the soundness of the operations on paper, but also that they subordinate their lives to the law of the large number. From the supra-individual perspective, an existential advantage was promised them: when in doubt, assault troops would be better off entering their own artillery fire from the creeping barrage, giving them cover, than in the sights of their enemy's machine guns, who would be given time to disappear into the creeping barrage and to reappear in front of them.[36] Through plans of action, the infantry did not merely lead up spatially to a creeping barrage, but also charged toward a temporarily lethal zone, for the innovation of releasing a preliminary barrage of dispersed gas over the battlefield before the main barrage went back to Bruchmüller with his fondness for war gas. This was gas that only in the best-case scenario had completely evaporated when, shortly after the bombardment, assault troops had drawn level with the target area.[37]

The eastern front has nearly fallen victim to collective forgetting, and only few specialist historians still remember that it became the first war theater of such tactical experiments and ultimately produced more corpses—and, above all, more nameless ones—than the western front.[38]

Perhaps the experience of these real horrors explains why the personal diary entries of Wittgenstein do not maintain the cool distance of a Martin Heidegger. Friedrich Kittler has demonstrated that Heidegger must have thought about assault tactics from the safe distance of the weather station on the western front. In *Being and Time*, the creeping barrage seems, in any case, to have served as a model for the fundamental existential motion of *Dasein*: "Anticipation [*Vorlaufen*: literally, running or moving forward] reveals to existence its extreme possibility as self-abandonment and thus shatters any clinging to an attained existence."[39]

In the postwar period, Wittgenstein—much to his friends' dismay—could imagine "what Heidegger means by being and anxiety. Man feels the urge to run up against the limits of language."[40] It remains to be elucidated what language means against this background.

During the defense against the Brusilov offensive, there was not enough time for Wittgenstein to entrust to his war diary even one line—apart from a quick prayer. For his first deployment as a reconnaissance officer, it was his mission to direct his own artillery fire to enemy positions. After a month of "colossal exertions," he has "thought much about all sorts of things, but strangely cannot make the connection to [his] mathematical thinking."[41] The next day he notes in his diary, "But the connection will be made! What cannot be said *cannot* be said!"[42] Here—and only here—the "double-entry bookkeeping" of the diary is broken. His private writings, otherwise encoded entirely according to the rules for reconnaissance officers, extend only here in plain text over both pages. This one time, the confessions, vows, expressions of despair and war experiences intertwine with the unencrypted philosophemes on the right side—from which the *Tractatus Logico-Philosophicus* will emerge. Thomas Macho's thesis is to be wholeheartedly embraced: "In many respects, the 'Tractatus' is the strangest war diary ever written."[43] Wittgenstein's later reflections also insist on the primacy of diagrammatic and cartographic constructs, on mathematical propositions, operations and orders, which are antecedent to thinking. The formation of his philosophemes is evidently characterized by the system of positional warfare—though this is much more fundamentally the case than Wittgenstein's biographers acknowledge when they merely register the appearance of war metaphors in his writings. At most, they seek to document what has ostensibly always already been a brilliant mode of rumination in its disturbance through war. If Wittgenstein's writings speak metaphorically of "laying siege"[44] to his mathematical and logical problems, of storming them and of "the blood" that he would rather pour before the fortress than "withdraw with nothing accomplished,"[45] the most striking aspect of this martial language is the anachronism—and Wittgenstein, before any contact with the front, was not the only one who stood under its spell. But the besieging and storming of fortresses and the heroic blood sacrifices are soon replaced by total sensual deprivation, death by asphyxiation, and trench systems in no man's land. The war that Wittgenstein—after he left behind the fortress city of Krakow—saw approaching from the observation post can no longer be decided by the taking of fortresses. It is necessary above all to capture spaces, because only their conquest promises a power advantage. Meanwhile, the conquered terrain appears more inhospitable after every offensive. Once the conquest of a considerable portion of land is successful, as in the Spring Offensive of 1918, the defense of the expanded front can then be one's undoing. If no metaphors of fortress structures can now be extracted from zones of visual deprivation, instruments of coordination can be—which Wittgenstein would find indispensable as professor of moral science at Cambridge University:

Language sets everyone the same traps; it is an immense network of well kept wrong turnings. And hence we see one person after another walking down the same paths & we know in advance the point at which they will branch off, at which they will walk straight on without noticing the turning, etc., etc. So what I should do is erect signposts at all the junctions where there are wrong turnings, to help people past the danger points.[46]

The fact that Wittgenstein orients himself only on the surface by everyday language and an everyday situation, but is secretly drawing on his military practices, can be gleaned from his files in the Vienna war archives:

During the battles at Ldziany [Wittgenstein] carried out his duties as a reconnaissance officer in an exemplary fashion. Staying at his post under the heaviest artillery fire, it was possible only thus for the battery to direct the fire at threatened points that the battery commander could not see. In this fashion sensitive losses were inflicted on the enemy at decisive moments.[47]

And ultimately the exact media transpositions that a reconnaissance officer had to perform are the basis of what will be called thinking in the most general sense in Cambridge's analytic philosophy lectures:

We can substitute a plan for words. And a thought may be a wish or an order. Truth and falsehood then consist in obedience or disobedience to orders. Thinking means operating with plans. . . . How do we know that someone had understood a plan or order? He can only show his understanding by translating it into other symbols. He may understand without obeying. But if he obeys he is again translating—i.e., by coordinating his action with symbols.[48]

The following description gives an impression of the coordinations (a word that was prevalent among artillery tacticians[49]), map operations, alphanumeric encryptions, and orders on the observation post: "Besides the captain, there is also a lieutenant there, a sergeant for operating the telephone and one for doing surveying work on the map, which is spread out over a board." The "serious military work" amounts to announcing "rapid and brief phrases . . . almost always in numbers," which remain in the memory of the civilian reporter merely as "an incomprehensible language."[50] Unlike the exciting noise of the barrage fire of his own side as well as the enemy artillery, the memory of the constant orders that he had to relay through the field telephone would remain abhorrent to Wittgenstein.[51] Nonetheless, on the observation post Wittgenstein finds himself for the first time in a position in which there is no question of the integrity of the chain of command. From the forward position, enemy trench mortars and artillery as well as "threatened points" of one's own line by the charging enemy infantry must be sighted with "scissors telescopes," coordinates must be determined on a battle map, and situation reports must be relayed to the battery commanders by telephone. Ultimately, the accurate fire by one's own artillery must be observed and, if necessary, must be ensured by further instructions. In the mediatic rule system of orders by telephone, answers, and indexical battle maps, the chain of command becomes a *circulus vitiosus*. There is no longer a supreme command, but only feedback loops and differences between what must be said and what must be shown, interrupted by lost telephone connections, deafening enemy fire, or the silence of the cannons, which reveal their position neither through noise nor through muzzle flash.

On the observation post, all instruments and media are geared to the handling of signs and orders. They become the fundamental activities that, for the philosophy of the twentieth century, still give rise to thinking. One instrument, however, is completely missing: There is no longer any mention here of the use of weapons. Even if artillery observers advance about as far as the assault troops in their initial position, and are thereby closer to them than to their own battery, a fundamental difference exists between the two. It emerges clearly from the accounts by Ernst Jünger, an early assault troop officer.[52] When Jünger is "given the job of observation officer"[53] due to a leg wound, he no longer has the front-line enemy positions in the sights of the assault rifle, but instead observes them with the telescope:

The observation post . . . was nothing more than a periscope through which I could view the familiar front. If the bombing was stepped up, there were coloured flares or anything else out of the ordinary, I was to inform the divisional command by telephone. . . .

The observation post was well camouflaged in the landscape. All that could be seen from outside was a narrow slit half hidden behind a grass knoll. Only chance shells ended up there, and, from my safe hiding-place, I was able to follow the activities of the individuals and units that I hadn't paid that much attention to when I myself had also been under fire. At times, and most of all at dawn and dusk, the landscape was not unlike a wide steppe inhabited by animals. Especially when floods of new arrivals were making for certain points that were regularly shelled, only suddenly to hurl themselves to the ground, or run away as fast as they could, I was put in mind of a natural scene. Such an impression was so strong because my function was a little like that of an antenna, I was a sort of advance sensory organ, detailed to observe calmly all that was happening before me, and inform the leadership. I really had little more to do than wait for the hour of the attack.[54]

In the coordination and synchronization of individual arms of the service, Ernst Jünger now realizes as an "advance sensory organ" the autopoetic as well as operational closure of the military body. Along with enemy movements, he also follows the combat units of his own side with an equally great identificatory interest. Only from the observation post do they reveal what must remain hidden at the operational level even from an assault troop officer. The "advance" organic-mediatic extension also shows operational limitations and dependencies in the cooperation of the individual arms. The fact that the disabled assault troop officer in the function of the observation officer now limits his activity to waiting for the hour of the attack expresses the fragmentation based on the real of the battlefield.

At the Somme, Jünger, physically recovered, yields to a reflex that reverses the decoupling of shooting and observation:

Later that morning, I was strolling along my line when I saw Lieutenant Pfaffendorf at a sentry post, directing the fire of a trench mortar by means of a periscope. Stepping up beside him, I spotted a British soldier breaking cover behind the third enemy line, the khaki uniform clearly visible against the sky. I grabbed the nearest sentry's rifle, set the sights to six hundred, aimed quickly, just in front of the man's head, and fired. He took another three steps, then collapsed on to his back, as though his legs had been taken away from him, flapped his arms once or twice, and rolled into a shell-crater, where through the binoculars we could see his brown sleeves shining for a long time yet.[55]

To carry the rifle on a loose strap ready to fire is a rule of the infantry tested in assault, to which Jünger himself no doubt contributed.[56] For the artillery observer, however, the exact opposite applies: he must avoid as much as possible any use of a weapon. The prime imperative is to maintain the secrecy of the observation post, because the enemy's defensive fire threatened to aim at it.[57] To hit the observation post also meant rendering inoperative the battery connected to it by telephone. Observation posts, which are withdrawn from sight through camouflage and do not directly deploy weapons, at least do not run the risk, as batteries do, that light measurement techniques home in on and precisely locate their muzzle flash from various positions or calculate their firing sites with the help of sonic measurement techniques. Judged by the systems that positional warfare produced, Jünger's shot is deplorable—for he endangers his own men more than the enemy's side. His demonstration of an aimed shot in fact celebrates an outdated minimal model of war: the duel; because not only does a "British soldier [break] cover behind the third enemy line" in the scene, but Jünger himself also strolls so ostentatiously to the sentry post that he does not even need to come out of cover in the first place.

When Jünger later, during the Second World War, in the middle of the Caucasus mountains, spots a scattered Russian unit on the opposite mountain ridge, the gaze into the telescope transports him into a lunar landscape, which can only mean the memory of the battlefields of the positional war, riddled with shell craters. With somnambulistic precision, a thought haunts him: "During the First World War one would still have opened fire on them."[58]

"From single shot to creeping barrage"[59]—thus the relevant literature sums up the irreversible course of history, propelled by a race between technology and tactics. The duel—which had been the principle to which, when in doubt, all complexities of war were still reduced up to the First World War—had thereby served its time. The Thirty Years' War might have produced the monopoly of violence, but the world war realized it in its totality. The threat of the death penalty in Prussian law did not put an end to the duel. Nor did Kant's appeal to reason, which argued that duelers by no means demonstrated the courage of the warrior, which was instrumental to states.[60] No state power was able to fight an institution that allowed the suffering and the exercise of violence for the restoration of honor. The duel is finally abandoned with the First World War, due to the intrinsic killing mechanisms of the world war, which revoke the equivalence with the duel. Even more than the right to exercise violence, the codex of putting one's own life at stake was rewritten. Dramatists such as Kleist with the Prince of Homburg could still invoke the freedom of death-defying courage, as a result of which every insubordination is forgotten. The positional war largely did away with a rank consciousness exemplified by the long-serving regiments of the death's-head hussars in favor of functional combat units, which answered to acronyms such as FEKA (*Fernkampfartillerie*, or "long range artillery") and risked the existence of their own units for the safeguarding of another in a circular logic per se and not only in a state of emergency. The war machine disavows the possibility of the individual seeking death for the defense of his own honor. Instead, it exposes life to many life-threatening risks in a quite particular fashion and under a specific directive.

The deep cut that logic makes in existence still speaks out of Wittgenstein's war diary: "If suicide is allowed then everything is allowed,"[61] it reads, consistent with the argument of axiomatic mathematics, according to which a contradiction is to be ruled out not because something false or untrue can be found in it, but rather because otherwise there would be a threat of the total indifference of all proof processes. Conversely, it can scarcely be an accident that Wittgenstein chooses the picture of a duel for the illustration of mathematical and semiotic contexts: "A is fencing with B."[62] Wittgenstein returns to the example several times in his writings—before the demand for a duel ultimately becomes a real option for coping with life.[63] Pairs of fighting men, Wittgenstein elaborates, can be represented only by extrapolating from one fighting pair to further pairs.[64] Thus, a relation between signs no longer necessarily refers back to the signified, but inherits other semiotic structures with their own logical relations. Between sign and signified a "logical identity" is not necessary, if internal— that is, not sayable, but showable—logical relations bring them together. For the production of identity with the signified, signs and modes of signification have to enter a complex that aligns their logical properties with the logic of the situations of the world. Propositions thereby do not simply describe situations of the world, but rather recreate them: "In a proposition a world is as it were put together experimentally. (As when in the law-court in Paris a motor-car accident is represented by means of dolls, etc.)"[65] On the eastern front, whenever he has enough time to do so, Wittgenstein sets himself the task of finding the connection between models, pictures, and "the signs on paper"[66] on the one hand and a "situation outside in the world" on the other.[67] Even if not all situations can be turned into "pictures on paper," Wittgenstein is certain that at least all "*logical* properties of situations" can be reproduced "in a two-dimensional script."[68] In a world whose situations are reflected without assistance in relations between signs and whose logic has to take care of itself, subjects ultimately have no place: "There is no such thing as the subject that thinks or entertains ideas."[69] It turns into an insurmountable limit, which cannot overcome itself in order to evaluate itself: "The Subject does not belong to the world: rather, it s a limit of the world."[70] Wittgenstein thereby encapsulates Jünger's "advance sense organ." On the observation post, which is withdrawn from the enemy's sight, and from which the battlefield experiences its limits through linguistic mediation, the "I of solipsism shrinks to an extensionless point and what remains is the reality co-ordinated with it."[71] In the solipsistic view, which according to a diary entry begins on the way into the firing position,[72] nothing "in the *visual field* allows you to infer that it is seen by an eye. For the form of the visual field is surely not like this"[73] (figure 6.1).

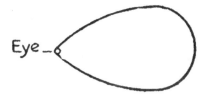

Figure 6.1
Wittgenstein's schema from his *Tractatus*.
Source: Wittgenstein 2001, 69.

Likewise, it cannot be concluded from the camouflaged battlefield that observing eyes are everywhere directed at it. The subject converges with a piece of paper and with a retina, which has abandoned thinking in duels but points the way for whole batteries.

Wittgenstein began his war diary two weeks after he entered the Austro-Hungarian Army as a volunteer with the sentence that "logic must take care of itself,"[74] so that "all we have to do is to look and see how it does it."[75] In the end, after the dissolution of the dual monarchy and with it the Austro-Hungarian Army, his work has "extended from the foundations of logic to the nature of the world."[76] If a problem is solved or a situation is managed, it loses its meaning. For Wittgenstein sentences are merely "ladders" that—as soon as they have proven their function—can be thrown away.[77] Jünger comes to the same conclusion in his study about the activity of the worker, whose type has emerged from the world war:

All of these concepts (*Gestalt*, type, organic construction, total) are *notabene* there by way of comprehension. We are not concerned with them as such. They can be forgotten or set aside without further ado after they have been used as magnitudes of work for the grasping of a definite reality which exists in spite of and beyond every concept; the reader has to see through the description as through an optical system.[78]

In war, even the most decisive technologies and tactics ultimately betray their design to the enemy, however effective they initially were, and demand their own surpassing. For this logic of surpassing, the war of the twentieth century is not a last resort of political clarification. Rather, this logic delineates the insurmountable playing field of a war game that constantly creates new unspeakable and inconceivable facts.

War on Mars: Wittgenstein's First Language Game

"What is the difference between language (M) [as mathematics] and a game? You might say: It ceases to be a game when things begin to become serious, and here seriousness means application."[79] Wittgenstein has returned to Vienna from Italian war captivity. He has again taken up his investigations of the foundation of mathematics—against the conclusion of his *Tractatus*, which declared the problem of logic solved once and for all and thereby showed how little that accomplished.[80]

In Wittgenstein's inquiries into the foundation of mathematics, war is again present. Now it has taken on the form of the war game, which outlasts every war and never runs out of material:

Think of the game of chess. Today we call it a game. Suppose, however, a war were waged in such a way that the troops fought one another on a field in the form of a chess-board and that whoever was mated had lost the war. The officers would be bending over a chessboard just as they now do over an ordnance map. Then chess would not be a game any longer; it would be a serious business.[81]

Wittgenstein does not recall a lived-through war, but puts himself on the level of the waging of it. The distance indicated by his analogy is double-.edged. To the extent that the battlefield no longer constitutes the tactical basis of war, but rather an increasingly detached level of symbolic configurations, to maintain distance from the battlefield is the very meaning of waging war.

With the game, the war game shares to the point of indistinguishability a sphere that avoids as far as possible hindering a free unfolding through facts and circumstances. However, in contrast to the mere game, the war game reserves for itself their transformation.

Wittgenstein suggests that chess was perhaps not always regarded as a game; it may well be a construction of the nineteenth century that the game could enter—to some extent phylogenetically—into a fundamental opposition to seriousness in the first place. Viewed ontogenetically, it is becoming increasingly apparent to Indologists that chess emerged from a war game: North Indian rulers of the sixth century moved terracotta figures over the sandy ground whose configuration resembled the four branches of their army.[82] Wittgenstein's chess analogy thereby implies an assumption and raises a dual question: the fact is that officers also operate on ordnance survey maps at the current moment, though there is no war—as Wittgenstein's use of the present tense makes clear. But what ensures that what is being conducted here is no mere game? And what ensures that in a game of chess no war is being waged? The answer to both questions is the same: nothing. It is precisely for this reason that Wittgenstein's philosophy of language games will be impervious to a critique that states that its very name manifests its irrelevance. Language games may certainly prove to be irrelevant, but no definitional power can anticipate such proofs. The limits of the game can only be played out, and thus with every war's end, the strategic and tactical playing through of a future one begins.

If a game ends up in an application, then it leads to a serious case, and the application leads to a scientific application, as Wittgenstein shows by way of the same analogy of the game of chess and the war game: "If on Mars there were human beings and they waged war against each other in the way chessmen do on a board, then their headquarters would use the rules of chess for prophesying. Then it would be a scientific question whether checkmate can be reached in a certain constellation of the game, whether mate can be reached in three moves, and so forth."[83] The difference between the application of signs and applications that derive from the application of signs is erased by Wittgenstein's analogy: the use of signs can be exhausted in the game as much as it can predict whole courses of war. Both extremes can be subject to one and the same "system of game rules"[84]—a system that does not fix the precise use of what it regulates.

On the basis of this dramatic indifference, Wittgenstein first approaches the subject of the game and links it to questions about the foundations of mathematics and its language. Commentators who regard the language game as an original philosopheme of Wittgenstein's overlook prolonged discussions in which reference is habitually made to sign games for the elucidation of mathematical foundations. Most recently and perhaps most impressively, Hermann Weyl had applied Hilbert's axiomatic-formalist proof procedure to concepts of the chess game and thereby provided Wittgenstein a basic schema for his language games.[85] Only when Wittgenstein transitions in his reflections from mathematically understood language games to general language games does his philosophy of mathematical language constructions experience its extension to a general ontology of grammatical rules.[86]

Wittgenstein's elaborations on the game, the war game, and mathematical formalizations were recorded by Friedrich Waismann so as to present them at the "Second Conference on the Epistemology of the Exact Sciences" in Königsberg.[87] At the conference, adherents of the logicist, formalist, and intuitionist schools convened to stake out once again their mathematical positions. At the beginning of the twentieth century, David Hilbert had committed mathematics to a formalism that, unlike Russell and Whitehead's logicism, would not persist in the attempt to base mathematics solely on logical elements. His proof procedures united arithmetical and logical operators in order to establish an axiomatic basis. Above all, the Dutch mathematician Brouwer and, in the early 1920s, Hermann

Weyl criticized Hilbert's mathematical operations, arguing that their claimed existence and effect could not be ensured through any mathematical intuition or constructive procedure.

At the time of the Königsberg conference, the dispute over the foundation of mathematics had already passed its peak. Weyl had cast his lot once again with Hilbert, for however ontologically questionable formalism might have appeared to him, he deferred to the success that the application of Hilbert's method within theoretical physics promised. Moreover, Kurt Gödel presented at the conference for the first time the basic features of his groundbreaking proof. Though this thwarted Hilbert's dream of securing a consistent mathematics, it was nonetheless demonstrated solely with Hilbert's formalist instruments and excluded Brouwer's intuitionism.[88] Even if Hilbert's long-term formalist objective of consistency and decidability proved to be unattainable, at least the formalist method had turned out due to Gödel's work to be unrivalled in measuring its own limits.

Brouwer had last participated in the discourse two years earlier with two talks in Vienna. For a long time they would remain his last public appearances—the quarrel with Hilbert had escalated, and it had for a long time no longer been only about mathematical entities.[89] Among Brouwer's listeners was Wittgenstein, who had first had to be persuaded by Waismann to attend the public event.[90] According to Herbert Feigl, who joined Waismann and Wittgenstein that evening, the lecture nonetheless impelled Wittgenstein to resume his philosophical ruminations.[91]

The question of whether Wittgenstein's position should be regarded as a further mathematical approach was at least addressed at the conference.[92] His standpoint was in any case understood to the effect that the meaning of a concept lies in its use.[93] It is thus clear that he proceeded primarily from Brouwer and Weyl's earlier attacks on formalist mathematics, which culminated in the idea that mathematics was more activity than theory.[94]

It is no longer the founders of the mathematical schools themselves, but meanwhile the generation of their successors who in Königsberg look back to the foundations of mathematics. Instead of Hilbert, John von Neumann spoke about formalism. Brouwer was represented by his pupil Arend Heyting, and Russell's logicist position was elaborated by Rudolf Carnap.

If, rather then asking about the mathematical constructs disputed in the foundational debate, one asks about a referential system that, as a precondition of the foundational debate, needs no introduction and is not doubted by any side, then the answer is "the game." Neumann, Heyting, and Gödel all took up at the conference in Königsberg the concept of the formula game; according to Heyting, the "word 'mathematics'" for the intuitionist means "a mental construction," for the formalist "a game with formulas,"[95] in which—to use Hilbert's words—a "technique of our thinking"[96] is first constituted. Neumann underscored in his contribution to the formalist explanation of mathematics that, though "the content of a classical mathematical sentence cannot always (i.e., generally) be finitely verified, the formal way in which we arrive at the sentence can be."[97] Therefore, it is less the statements themselves that should be investigated and more the methods of proof, which should be understood as a "combinatorial game played with the primitive symbols."[98] In the run-up to the conference, Wittgenstein seems to comment in advance on Neumann's elaborations:

Something in formalism is right and something is wrong. The truth in formalism is that every syntax can be understood as a system of game rules. I have thought about what Weyl may mean when he says that the formalist conceives of the axioms of mathematics as like the rules of a chess game. I want to say: Not only the axioms of mathematics but all syntax is arbitrary.[99]

That axioms dispense with substantiation was mathematical consensus: if axioms are to constitute the basis of all derivations, they themselves escape every derivation. But Neumann and—under a different sign—Wittgenstein now see proof procedures too as springing from the arbitrariness of a semiotic game.[100] This viewpoint also dispels all the origin legends that ultimately regarded proof figures as emerging from the genius and the inspiration of mathematicians.

In any case, Neumann disagreed vehemently with Carnap and insisted that "it is actually meaningless symbols that are introduced. But for Hilbert the introduction of these meaningless symbols is not an end in itself."[101] Neumann was the first at the conference to recognize the consequences of Gödel's proof, and he built a bridge for him with his statement. Gödel's commentary in the conference publication, which once again discusses his proof of the impossibility of an irrevocable consistency in mathematics, takes up Neumann's characterization of formalism and even adopts his conceptualizations. Formalism, according to Gödel, is about a "purely combinatorial property of certain sign systems and the 'game rules' that apply to them," with which "combinatorial facts [can ultimately be] expressed in the symbols of mathematical systems."[102]

The game with formulas was thus by no means abolished with the waning of the foundational debate. On the contrary: there is no method that can demonstrate possible contradictions of combinatorial facts except for these themselves. Gödel could also declare the game with signs a combinatorial fact because it had long been common among mathematicians to plumb the foundation or the groundlessness of mathematics on this basis. However diverse the mathematical standpoints of Gottlob Frege, Weyl, Brouwer, Hilbert, or Bernays appeared in the foundational debate, what they all have in common is that they determine the nature of mathematics in the difference with or in the correspondence to the game.[103] Thus it comes as no surprise that the first historians of mathematics, such as Oskar Becker and Jean Dieudonné, described Hilbert's formalism as a game with formulas.[104] If mathematicians thus began around 1930 to talk about the game, then it was no longer necessarily connected with the pejorative sense that had often been attached to the term in the past—for example, when Gauß had still spoken of the "meaningless formula game,"[105] or when even Hilbert himself still admonished in a lecture in 1919 that mathematics was "not like a game" in which "the tasks are determined by arbitrarily conceived rules."[106] But by taking up the game as a sign system, the mathematical discourse experiences a radical expansion of its playing fields.

Sign Game

The first person after Leibniz to investigate games seriously in terms of their mathematical efficacy was Paul Du Bois-Reymond. Unlike his brother Émile, who more than almost anyone else was responsible for physiology's rise to the status of a leading science in the late nineteenth century, Paul attended, alongside his medical studies, Dirichlet's lectures on the integration of partial differential equations and ultimately held professorships in pure and applied mathematics.

The fact that a caesura within mathematics is nonetheless associated less with Paul Du Bois-Reymond and more with his brother is due to the altered position of mathematics with respect to other disciplines.

"Ignoramus et ignorabimus" were Émile Du Bois-Reymond's concluding words at the Leibniz meeting at the Academy of Sciences in 1880: motive forces and consciousness are transcendentally unfathomable.[107] "In mathematics there is no ignorabimus!"[108]—with these words Hilbert opened the Second International Congress of Mathematicians in Paris in 1900. Thirty years later, he still had not grown tired of repeating for radio listeners: "We must know, we shall know."[109] When Hilbert presented at the congress the fundamental program of the dawning century, he rejected a scholarly class to which the brothers Du Bois-Reymond still belonged. Hilbert's list of twenty-three problems that he first posed to the congress participants—and, to some degree, to mathematicians to this day—may have caused a stir in part because it broke with the humility of an auxiliary science that waited for the annual offering of a prize question from the academies. Hilbert replaced the monolithically embodied knowledge that academies spread out in their departments with the mathematical operation of the mathematical institute.[110]

His program was not merely interested in the general solvability of mathematical problems, but in the development of the axiomatic method to a procedure of knowledge that ultimately no science could circumvent.[111] To Émile Du Bois-Reymond's Latin maxim, which emphasized limits of knowledge for the living metaphysical body, it was not even necessary for Hilbert's positivism—propounded in East Prussian dialect and primarily directed toward sign systems—to respond. However, Du Bois-Reymond's doubts were already coming from a level that had long been subject to the criteria of mathematics. He had vehemently advocated for the increased inclusion of mathematics in the curricula of the humanistic *Gymnasium*—if necessary, to the detriment of the ancient languages. The only answer his brother Paul Du Bois-Reymond and Hilbert had for the question of what mathematics could be applied to was the equally positive counterquestion: "What is not [applied] mathematics?"[112]

But the decisive, epochal caesura around 1900 is not to be found in different views as to how material foundations of force and of consciousness or life can be mathematically grasped. Rather, it is manifested in the struggle to trace mathematics back to foundations other than the Platonic heaven of ideas. Upon closer inspection, Hilbert's program is therefore directed less against Émile Du Bois-Reymond than against his mathematician brother Paul, who was the first to proceed from the possibility of alternative foundations and not merely from systems in mathematics.[113]

In the course of this, Paul Du Bois-Reymond had in 1882 already called formalism by its name and declared it dead—long before Hilbert elevated a formalist mathematics based on axioms to a program. Du-Bois Reymond stated:

A purely formalist-literal structure of analysis, which is what the separation of number and sign from quantity amounts to, would ultimately degrade this science, which is in truth a natural science, even if it only draws the most general properties of the perceived into the domain of its research, to a mere game of signs, in which arbitrary meanings would be attached to the written signs as to chess figures and playing cards. As delightful as such a game can be, as useful for analytic purposes as the solution of the task of pursuing the rules between the signs, which emerged from the idea of quantities, to their ultimate formal consequences may even turn out to be, this literal mathematics, if it were completely detached from the ground on which it grew, would nonetheless soon enough exhaust itself in unfruitful sprouts, while the science that *Gauss* [italicized in original] so truly and profoundly called *Grössenlehre* [theory of quantities] possesses in the natural domain of human perception, which is always expanding, an inexhaustible source of new research objects and fruitful stimulations. Without question, with the help of so-called axioms, of conventions, ad hoc philosophemes, inconceivable extensions of originally clear concepts, one will be able to construct retroactively a system of arithmetic resembling in all respects that which emerged from the concept of quantity, in order to cordon off calculative mathematics, so to speak, through dogmas and defensive definitions from the psychological domain. An extraordinary acumen can even be applied to such constructions. Moreover, one would be able to think up in the same fashion other arithmetical systems, as has happened. The ordinary arithmetic is precisely the only one that corresponds to the linear concept of quantity, is so to speak its first registration, while analysis constitutes its highest development with the limit concept at the forefront. Even the difficulties of the limit concept, which we will readily confront fearlessly, one may believe that one can solve through symbolics. It will scarcely succeed. For every analyst who is more than a combinatorialist will want to pursue the origin of the game of signs, and thus find himself once again facing the circumvented problems.[114]

Du Bois-Reymond thus demands that rather than cordoning off the "psychological domain" through "dogmas and defensive definitions," one must draw "real quantities" from a "domain of human perception, which is always expanding." There are reasons external to mathematics for the fact that even a proof of the existence of a limit to continuous sequences, which proceeds with the help of discrete decimal fractions, does not become a paradox. It is not discrete signs that first segment the continuum into discrete quantitative sequences, but the "peculiarity of thinking" itself, which has its external sign in the "sight perception" of the "jerky turning of the eyeball."[115] Du Bois-Reymond, who initially made his mark with studies of the blind spot,[116] placed mathematics on foundations that remain bound to human psychology and physiology. Furthermore, he distinguished real quantities from mathematical ones—each of which he differentiated in turn. Mathematical quantities, however, bring together combinatorial and "logical processes."[117] Du Bois-Reymond would refer them entirely to the domain of the mind, if another "combinatorial area" did not come to the fore—that of the game: "It cannot be denied that the knight's move problems, but especially the so-called endgames of chess . . . exhibit the character of genuinely mathematical tasks, only within an extremely limited combinatorial area."[118] From a field that in the late nineteenth century was otherwise perceived only as mathematics for amusement, Du Bois-Reymond extracts a new dimension—only to reject it immediately. The "game quantities" have "unreality in common" with mathematical quantities—without, however, standing like the latter in "close relation" to reality.[119]

Proof Figures beyond the Sovereign Subject

The reference to the game within mathematics, although or precisely because it appears peripherally, is the only one that persists as a "common platform of all discussions," when Hilbert, as Weyl summed up laconically, "postulated his proof theory," and toppled mathematics as a "system of substantial, meaningful, insightful truths."[120] Though Du Bois-Reymond had found in "the game of signs" a "formalist-literal structure," it did not come into serious consideration as a mathematical foundation due to its "limited combinatorial area." Brouwer and Weyl's critique of Hilbert's program of formalization was diametrically opposed; formalism practices a formula game that goes beyond domains still accessible through acts of thinking in Brouwer's sense.

Instead of proceeding merely from different phases in the confrontation over mathematical foundations,[121] which have to manage the increasingly reciprocal references of their mathematical structures, it is the references themselves that must be interrogated. These cannot be constrained even by the rigid referential system of axiomatic postulations. It is here that the caesura and the fault line first become apparent that overtook the discourse of mathematical foundations. And this pertains above all to the mathematical sign, which is now taken as at once object and foundation:

No more than any other science can mathematics be founded by logic alone; rather, as a condition for the use of logical inferences and the performance of logical operations, something must already be given to us in our faculty of presentation [*in der Vorstellung*], certain extralogical concrete objects that are intuitively [*anschaulich*] present as immediate experience prior to all thought. If logical inference is to be reliable, it must be possible to survey these objects completely in all their parts, and the fact that they occur, that they differ from one another, and that they follow each other, or are concatenated, is immediately given intuitively, together with the objects, as something that neither can be reduced to anything else nor requires reduction. This is the basic philosophical position that I regard as requisite for mathematics and, in general, for all scientific thinking, understanding, and communication.

And in mathematics, in particular, what we consider is the concrete signs themselves, whose shape, according to the conception we have adopted, is immediately clear and recognizable. This is the very least that must be presupposed; no scientific thinker can dispense with it, and therefore everyone must maintain it, consciously or not.[122]

What scientific thinkers have to envision consciously—and everyone else comprehend unconsciously—are signs, which do not point beyond themselves as referents. The assertion that "number signs [*Zahlzeichen*, or numerals], which are numbers and which completely make up numbers" nonetheless become the sole object of consideration, "but otherwise have no meaning at all," provokes the first critical inquiries: "Can there be a sign without meaning?" Aloys Müller, who posed this question to Hilbert, also made an attempt to answer it himself:

A sign always *signifies* something *that is different from the sign itself*. Sign and signified object are *assigned* to each other. . . . If Herr Hilbert wants to insist that 1 and + are without meaning, then they are not signs, but in that case merely drawings [or] figures. . . . Is that a foundation for number theory? Certainly not. With the necessary imagination, one thus gets pretty moldings or wallpaper borders and for each a manufacturer's trademark, but not mathematics.[123]

Hilbert left the defense against the critique to Paul Bernays, who could not help making linguistic concessions. Instead of signs, it would be better to speak in the future of figures[124]—instead of numbers, numerals.[125] However, the Hilbert school proved to be firmer in the matter. Where the border ran between "meaningless figures"[126] and signs established by them remained subject to debate.

Hilbert seeks to solve the paradox of representing infinity and continuity with finite and discrete signs by declaring them solely a matter of axiomatic postulations and thus of sign systems. In contrast to Du Bois-Reymond and all mathematicians who, with Leibniz, proceeded from the assumption that the world makes no leaps, Hilbert excluded infinity from the physical world:

For everywhere there are only finite things. There is no infinite speed, and no force or effect that propagates itself infinitely fast. Moreover the effect itself is of a discrete nature and exists only in quanta. There is absolutely nothing continuous that can be divided infinitely often. Even light has atomic structure, just like the quanta of action. I firmly believe that even space is only of finite extent, and one day astronomers will be able to tell us how many kilometers long, high and broad it is. And although there are in reality often cases of very large numbers (for instance, the distance of the stars in kilometers, or the number of essentially different games of chess) nevertheless endlessness or infinity, because it is the negation of a condition that prevails everywhere, is a gigantic abstraction—practicable only through the conscious or unconscious application of the axiomatic method. The conception of the infinite, which I have grounded through detailed investigations, answers a number of important questions; in particular, it shows the baselessness of the Kantian antinomies of space and of the unlimited possibilities of division, and thus of the difficulties that crop up thereby.[127]

Nothing remained of Du Bois-Reymond's mathematical foundations, which in the final analysis were always perceived from limits and emanated from perceiving bodies. At most, gestalt-theoretical considerations still determine the discussion of the postwar years. In the foundational debate, bodies no longer matter, but space certainly matters everywhere.

That the body, understood psychologically, is excluded from a mathematical framework may still be traceable to a disciplinary differentiation. The legitimation strategy within mathematics to demonstrate its use for other disciplines and thereby its own necessity becomes dispensable. But Hilbert's formalist mathematics itself still reaches beyond its own discourse by charging the general concepts of its choice in a specific fashion.

The concept of the body is not suited to that strategy. It can be the object of mathematical procedures that prove its calculability, but its investigation is left to other disciplines. Space is an entirely different matter—it stands at the center of the foundational debate. The mathematics of the postwar period asserted over all other disciplines the claim to a genuine concept of space and set out to provide means and techniques for the mastery of it. In this, it is crucial to free the Kantian "*a priori* theory" of the last "anthropomorphic dross."[128] Hilbert's formally cultivated mathematics as much as its intuitionist version demand an unmediated access to elements constitutive of space. While formalism dispenses with phenomenological interpretations[129] and abandons itself to calculation as an object disseminated in signs, intuitionism subordinates a spatial continuum to the primacy of time. For Brouwer, mathematics therefore has to merge completely with its activity; for Hilbert, on the other hand, it exists entirely on paper: "The question of where scientific exactness exists is answered differently by the two parties: the intuitionist says: in the human intellect, the formalist: on paper."[130] The debate over the foundation of mathematical operations has an underlying geopolitical subtext. Brouwer is chiefly responsible for introducing it into the discourse. Before foundational mathematical publications, he wrote about early mythical times and their disenchantment through techniques of land reclamation:

Holland was created and was kept in existence by the sedimentation of the great rivers. There was a natural balance of dunes and deltas, of tides and drainage. Temporary flooding of certain areas of the delta was a part of that balance. And in this land could live and thrive a strong branch of the human race.

But people were not satisfied; in order to regulate or prevent flooding they built dykes along the rivers; they changed the course of rivers to improve drainage or to facilitate travel by water, and they cut down forests. No wonder the subtle balance of Holland became disturbed; the Zuyder Zee was eaten away and the dunes slowly but relentlessly destroyed. No wonder that nowadays even stronger measures and ever more work are needed to save the country from total destruction.[131]

His dissertation on the foundation of mathematics soon ignited a quarrel with his doctoral adviser,[132] who rejected large portions of his work and cut the sentence stating that science serves human beings solely in the struggle against their own kind and against nature—that it ultimately has only the value of a weapon.[133] Even astronomical models, he argued, were subject to the will of individuals. They knew how to read from the measuring instruments those values that lent themselves to the construction of theories. Brouwer therefore declared: "The laws of astronomy are no more than the laws of our measuring instruments."[134] For Brouwer, mathematics—looking backward beyond the contrary positions of Kant and Leibniz—had to find its way back to a mystically charged primal intuition.[135]

All the more surprising is the turn in Brouwer's career when, against the will of the professor of applied mathematics, the latter actually became a weapon for him. Perhaps to avoid serving the Dutch Army once again as a reservist—his first period of service must have been traumatic—from 1915 on, Brouwer took the bull by the horns: he began to delve into photogrammetry and submitted a memorandum to the ministry for defense.[136] 1915 was the year he learned from Schönflies, during a visit to Göttingen, that many of the young mathematicians there were used by the military for measurements and transformation calculations of aerial photographs—and members of the Academy directly advised the General Staff.[137] But 1915 is also the year a new era dawned for photogrammetry in general. At that time, Oskar Messter registered for a patent for his "Method for Producing Photographic Images from an Aircraft."[138] Aviatics and photography could thus enter a media-technological alliance that created maps out of survey photographs in a continuous technological processing chain—that is, with the exclusion of human perception. Brouwer calculated for the Dutch General Staff how, thanks to trigonometric methods, unavoidable angle differences in serial photographs of terrain could be reconciled and topographical maps with a larger scale than what was previously common could be created.[139] But the chief of the General Staff declined and declared the basic scale used at that time to be sufficient. The works on photogrammetry remained without any positive response and Brouwer fell into a depression.

When Weyl, after the war's end, adopted Brouwer's standpoint on mathematical foundations, geopolitical resonances were in play from the beginning. In this, Hilbert anticipated Weyl when, in a lecture on axiomatic thinking in Switzerland in 1917, he drew an analogy between the "life of science" and that of states, which "have to be well ordered" not only in themselves, but also in their relations to each other.[140] For Weyl, the situation culminated more drastically, resembling the "separation of the Occident from the Orient" at the time of the Persian wars, the tension and overcoming of which "became the driving motive of knowledge for the Greeks."[141] Against this background, the "antimonies of set theory" are

regarded as border conflicts that concern only the most remote provinces of the mathematical empire and can in no way endanger the inner solidity and security of the empire itself, its actual core areas. . . . Indeed: any serious and honest reflection must lead to the recognition that those detrimental effects in the border regions of mathematics must be judged as symptoms; in them comes to light what is hidden by the outwardly shining and frictionless operation in the center: the inner instability of the foundations on which the construction of the empire rests.[142]

Ultimately, all that remains for Weyl is to "gain solid ground" in the face of the "looming dissolution of the polity of analysis" and to declare "Brouwer—that is the revolution!"[143] Subsequently, Hilbert accused Weyl and Brouwer of an "attempted coup," a "dictatorship of prohibitions," and "terror."[144] There is no question that the use of a rhetoric for the continuation of war by verbal means may say as much about the time after 1918–1919 as about the intensity of the mathematical foundational debate. But the fact that, for all the metaphors, effective methods for the calculation of spaces and borders were flourishing should not be overlooked.

Just as proof figures come to the fore in the mathematical discourses, the metaphorical recedes. Bringing the measurement of a natural space under control is now less urgent than sketching spaces that arise from a sign-based apparatus and that are anything but mathematically secured.

From an operational point of view, fundamental questions of the establishment of a "mathematical apparatus"[145] and of a state apparatus are now on the same page. Brouwer, for one, recalls that it was "not just theoretical sciences like paleontology or cosmogony" that depended on trust in the principles of classical logic incriminated by him, "but also governmental institutions like the rules of procedure for a criminal trial."[146] He calls into question the idea that a theorem or a formula should be regarded as true merely because a proof proceeding from the opposite assumption leads to a contradiction. Conversely, a theory may not be false merely because the assumption of an opposing proof reveals no contradiction. To lend his argument weight, he compares mathematical procedures with those of criminal law: "[An] incorrect theory remains incorrect even if it cannot be disproved by contradiction, in the same way that a criminal policy remains criminal even if it cannot be condemned and stopped by any legal process."[147] Weyl too turns to the comparison between mathematical and police state methods. Thus, one should "beware of the idea that, when an infinite set is defined, its elements are, so to speak, merely spread out before one's eyes, and that one need only go through them in succession, as a police officer goes through his register, in order to find out whether in the set an element of this or that sort exists. With respect to an infinite set, that is senseless."[148] Brouwer and Weyl clearly no longer adhere to Hilbert's textbook-like example sentences, which exclude any reference to the present, such as the still rather scholastic: "Aristides is corruptible." In contrast, to demonstrate to formalized mathematics its lack of constructive methods for the securing of the infinite by finite means, Brouwer and Weyl adhere to state organs, the legal constitution of which was no less contentious at that time. Brouwer implies that when the criminal facts of a case apply to the executive himself, this does not challenge the legislative power. And Weyl proceeds from the assumption that the police operate with a register that encompasses a well-defined and clearly circumscribed set. Indeed, with their analogies they touch on areas of the state constitution that could not have been more controversial among constitutional law scholars of their time. Carl Schmitt, whose dominance in constitutional law discourse arose with the final days of Weimar, would not have agreed with Brouwer that the law is still in force when the executive is not in accordance with it. On the contrary, for Schmitt, "authority proves that to produce law it need not be based on law."[149] The ultimately decisive exceptional case "makes relevant the subject of sovereignty, that is, the whole question of sovereignty. The precise details of an emergency cannot be anticipated, nor can one spell out [*Aufzählen*] what may take place in such a case, especially when it is truly a matter of extreme emergency and how it is to be eliminated."[150] Jurists may still speak colloquially of enumeration (*Aufzählen*), but formalists have long required several words: they know how to differentiate countably (*abzählbar*) finite sets from uncountably (*überabzählbar*) infinite ones—and these, in turn, from counting to transfinite numbers.

In mathematics, it has been known since Georg Cantor's set theory that there are infinite sets of different power or size (*Mächtigkeit*). At least the largest set of real numbers can therefore not be deterministically circumscribed by a system of order. The problem that the sovereign subject stands outside "the normally valid legal system" and "nevertheless belongs to it," for it gives him the right to suspend it,[151] doubles a situation that already applies to formal systems as such. And Weyl, as shall be shown in greater detail, proclaims a revolution for the very reason that he, with Brouwer, does not proceed solely from numerical sequences "determined by law" but also from those that emerge "from step to step through free acts of choice."[152]

These discourses of jurists and mathematicians correlate also with respect to the formation of their factions. Schmitt's definition of the sovereign who decides on the state of exception opposes the purely formalist legal theory of his rival Hans Kelsen. Schmitt argues against Kelsen that the case that must first come to pass in order to be able to be judged and decided escapes the possibility of regulation by normative legislation. On the opposite end of the political spectrum, Evgeny Pashukanis—a leading representative of Marxist legal theory—ranked alongside Schmitt as one of Kelsen's critics. He viewed Kelsen's neo-Kantian legal positivism as a "legality of ought purified of all psychological and sociological 'residues,'" which

neither has nor can have any rational definition at all. . . . On the level of the juristic ought there is only a transition from one norm to another on the rungs of a hierarchical ladder, on the top rung of which is the all-encompassing, highest norm-setting authority—a limit concept from which jurisprudence proceeds as from something given. . . . Such a general theory of law, which explains nothing, which turns its back from the outset on the facts of reality, that is, of social life, and busies itself with norms, without taking an interest in their origin (a meta-juridical question!) or in their relation to any material matters, can certainly at most lay claim to the name theory in the sense in which one, for example, customarily speaks of a theory of chess.[153]

Pashukanis seems to have taken up the reservations that Brouwer and Weyl already expressed with respect to formalist mathematics and directed them against Kelsen's formalist legal theory—not least of all, his choice of words and the chess analogy suggest as much. Weyl certainly lent a hand to the discursive leap by himself attaching a political, police function to simple quantifiers of predicate logic: "The 'there is' arrests us in *being* and *law*, the 'every' releases us into *becoming* and *freedom*."[154]

Since Cantor and Frege's contributions to set theory, mathematical conceptions of space are no longer dominated by the limit value problem of analysis, but by possibilities of ordering that belong to the measuring and counting numerical sequences and sets themselves. Weyl shares this aim when he designs a continuum "into which the individual real numbers undoubtedly fall, but which by no means dissolves into a set of real numbers as finished beings," but rather offers "a medium of free becoming."[155]

Du Bois-Reymond had already speculated about sequences that escaped every law, because they were either based on rolls of the dice or broke free of "human company" and continued "independently on the way into the endless," through "a fixed rule" that was "given to them for the journey." Ultimately, Du-Bois Reymond distanced himself with an empirical view of mathematics from "assuming and weaving into the mathematical thought process things of which we have and can have no conception."[156]

Weyl, on the other hand, inspired by Brouwer and supported by his own investigations of recursive sequences of dual fractions, was serious about the design of "judgment instructions" that are "self-sufficient" and "even contain at their core an infinite abundance of real judgments." The clearly delineated judgment instructions "formulate," according to Weyl, "the justification for all the singular judgments to be 'redeemed' from them." The judgments themselves cannot be "redeemed" except through the execution of the process that produces them: "This happens insofar as we allow the emerging [*werdende*, i.e., becoming] sequence of choices at every step to produce a number or nothing *or cause the breaking-off of the process, its own death* and the annihilation of its previous production."[157] The continuum, and with it space, now appears to Weyl to be "becoming toward within into the infinite." He opens up a space that corresponds in fundamental respects to Carl Schmitt's design of an "order of large spaces" (*Großraumordnung*) in international law. What is significant in Schmitt's *Völkerrechtliche Großraumordnung* (his notorious text on the large spatial order) is first and foremost the supplanting of the juristic concept of the *Reich*—the realm or empire—by that of space. At the beginning and conclusion of the text, Schmitt does not neglect to deal with the concept "large space" (*Großraum*). Though *Raum*, or "space," still has a mathematical-physical sense, *Großraum* encompasses more of a "technical-industrial-economic-organizational domain."[158] It should no longer be regarded as mathematically neutral space, in the emptiness of which "the perceiving subject" inscribes "the objects of its perception," in order to "localize" them.[159] With *Großraum*, Schmitt seeks to leave behind the "mathematical-scientific-neutral field of meaning" that still adheres to the concept of "space": "Instead of an empty dimension of surfaces and depths in which physical objects move, the cohesive *space of achievement (Leistungsraum)* appears, as is proper for a history-filled and historic *Reich*, which brings along with it and bears within itself its own space, its inner measures and borders."[160]

Schmitt first comes to this conclusion in 1941. He formulates it in a chapter that he adds to his text on *Großraumordnung*, which appeared in 1939. It is in this chapter that Schmitt first sees the natural sciences open up to the *Großraum* through Max Planck's "World Picture of New Physics" and Victor von Weizsäcker's biological "space of achievement" (*Leistungsraum*). However, neither the experience of the Blitzkrieg nor the mediation through the natural sciences was necessary for his concept of space. In the medium of paper, which was itself subject to debate as the mathematical foundation, the mathematicians in Göttingen, Amsterdam, and Zürich had already expanded on orders and arrangements of space for a long time. For all the criticism of Schmitt in the evaluation of the consequences of his juristic and political engagements, the question of sources that he intentionally fails to acknowledge has remained unaddressed. His concept of "large space" is clearly not merely motivated by a political movement, but is also shaped by a mathematical discourse that had already developed on its own a political surplus. Thus Schmitt still clings to the existence of a sovereign subject that decides on a history-filled space, while mathematicians had long since moved on to studying the condition of possibility of a medium of free becoming, in which decision-making power emanates from the contingency of a game of signs.

The Shared Origins of Game Theory and the Universal Machine

Sign games, which set their own unfolding in motion, leave behind naturalistic and life-philosophical conceptions. Henri Bergson, for one, still found it completely unthinkable that mathematical symbols could designate duration and movement otherwise than only indirectly. They themselves are intrinsically immobile: "For the geometer all movement is relative: which signifies only, in our view, that none of our mathematical symbols can express the fact that it is the moving body which is in motion rather than the axes or the points to which it is referred."[161]

For Bergson, the mathematical symbol that directly designates a body in motion would have to coincide with it. But the fact that symbols have always already been set in motion in games—such as the sign-bearing game pieces of the medieval Battle of Numbers, movable type in book printing, and the "types" in Reiswitz's tactical war game, on instrument displays and in calculating machines—does not occur to Bergson. Most likely, he would scarcely have seen in these examples anything but exceptions in which the categories are mixed in a confused fashion.

In fact, mathematicians in the twentieth century began to design symbol systems with operations that, at least according to the fiction, are accompanied by movements. In this, it was not a matter of representing life-world phenomena of movement in imagined or actually constructed mobile sign configurations. Rather, it was the reverse: a matter of exploiting new possibilities of making-calculable through arrangements of signs set in motion.

If one disregards Adam Smith's "imaginary machines," which promised to transfer the principle of Newton's celestial mechanics to state constitutions,[162] then the eugenicist and statistician Francis Galton, toward the end of the nineteenth century, was the first to begin consistently representing inherently abstract mathematical models in machine models and thereby reversing the classical process of analysis. Galton initially furthered developments in representing statistics through graphic methods. Nonetheless, behind his efforts in the visualization of inherently abstract characteristics through mechanisms was an elaborate psychology of imagination that he called "mental imagery."[163] This psychology would also enable the introspective design of "mechanical illustrations."[164] There is no evidence that Galton ever mechanically implemented his illustrations as a model. Therefore it would be false to assume that his illustrations are construction drawings; rather, they must already be convincing as fictive machines. It is no longer necessarily the task of mathematics to calculate specific properties of mechanisms or machines. Rather, the latter now serve mathematics. They are therefore released from every other purpose. And the imagination too thus receives a different status. In contrast to the fictionalism of thought experiments, which also proceed from counterfactual assumptions, it is not compulsorily fictive. On the contrary—no imaginary surplus presses for the illumination of a domain that would pose a problem in its formal development. Rather, only what can easily be technically realized is imagined and transposed into the fictional. If the mechanical realization presented a problem, the fiction should not be maintained. If the realization is already established in the fiction, every effort of realization also ultimately becomes unnecessary. Galton's mechanical illustrations therefore bring to light a new discursive figure with quite far-reaching consequences. In contrast to an only verbalized hypothesis, they show an effective way to implement what is said operationally.

Alongside game configurations, the invocation of fictive machines also plays an important role in the twentieth century during the mathematical foundational debate between the formalist and intuitionist camps.[165] In a controversial fashion, they reveal limits that can be symbolized and formalized on the one hand and imagined and thought on the other. Ultimately, the dispute was not least of all about whether in mathematics a shift onto a metalevel would be possible. While David Hilbert elevated this very shift into a program and hoped from a metalanguage that it could precisely describe an operationally occurring symbolic language, E. J. Brouwer raised considerable doubts. As an insuperable epistemic problem, he invoked temporality, which is proper and essential to a mathematical operation and which for him is realized through an act of thought. For Brouwer, this fundamental aspect is irretrievably lost in the attempt to refer to the operation in retrospect—not least of all because the reference is, in turn, subject to its own temporality.[166]

In the invocation of fictive machines, there is however no discernible platform that bridges the splintering mathematical discourse of the foundational crisis and keeps it in productive motion. In contrast to a fixed and fixing metalanguage, fictive machines abolish the temporal-performative structure called for by the intuitionist Brouwer and viewed by the formalist Hilbert as a temporary operation that challenged the mathematician to catch up to it in thought. The foundations of mathematics conceived in the form of fictive machines and models and as sign games appear, however, without being introduced systematically or conceptualized rigorously. Still, fictive machines should by no means be regarded as mere examples. Rather, they form discursive levels that regularly come into play.

Someone who perceived from up close the "technization of formal-mathematical thinking"[167] was Edmund Husserl: "[A] technization takes over all other methods belonging to natural science. It is not only that these methods are later 'mechanized.' To the essence of all method belongs the tendency to superficialize itself in accord with technization."[168] His *Crisis of European Sciences*—written in the 1930s, it should be recalled—does not primarily seek the reasons for "the crisis of our culture"[169] in the humanistic sciences—to question their scientific nature was not new.

Rather, he discerns drastic "shifts in meaning"[170] specifically in the prosperous positive sciences, above all in pure mathematics.[171] Husserl's *Crisis*—in spite of his and the general political critical situation—evidently still stands under the spell of the mathematical foundational crisis.[172] For him, science turns into *techne* not because thought first pushes toward mechanization but rather because it already thinks of itself as mechanics:

Are science and its method not like a machine, reliable in accomplishing obviously very useful things, a machine everyone can learn to operate correctly without in the least understanding the inner possibility and necessity of this sort of accomplishment? But was geometry, was science, capable of being designed in advance, like a machine, without an understanding which was, in a similar sense, complete—scientific? Does this not lead to a *regressus in infinitum*?[173]

During Husserl's lifetime, a text already appeared that would have caused his rhetorical questions to waver in a fundamental fashion—had this text initially received more than the attention of only a very small circle of mathematicians. In 1936, two years before his death and in the middle of the period in which he wrote *Crisis*, the British mathematician Alan Turing, at King's College in Cambridge, produced his now famous text "On Computable Numbers with an Application to the Entscheidungsproblem."[174] The fact that Turing in this text performed a decisive proof on the question of the formalizability of decidability by means of a fictive machine, a paper machine,[175] went beyond the previously prevailing, supplemental meaning of fictive machine constructs. Turing's machine fiction serves in this text not as an exemplification of his proof; it *is* the proof.

For Husserl, this turn was not foreseeable; his gaze was still directed entirely at the praxis of a thoroughly axiomatized formal mathematics:

But now [only] those modes of thought, those types of clarity which are indispensable for a technique as such, are in action. One operates with letters and with signs for connections and relations (+, ×, =, etc.), according to *the rules of the game* for arranging them together in a way not essentially different, in fact, from a game of cards or chess.[176]

Husserl denies an epistemologically open horizon to a formal mathematics conducted in this fashion and he grants it only the status of a "mere art." [177] The idealizations and constructions methodically practiced "intersubjectively in a community" can be used "habitually and can always be applied to something new—an infinite and yet self-enclosed world of ideal objects as a field for study." Mathematical symbols are thus conceived as being objectively knowable and available without requiring that the formulation of their meaning be repeatedly and explicitly renewed. On the basis of sensible embodiment, e.g., in speech and writing, they are simply apperceptively grasped and dealt with in our operations. Sensible "models" function in a similar way, including especially the drawings on paper which are constantly used during work.[178]

Ultimately, mathematical symbols would therefore correspond to tools such as pliers and drills, created only for "mental manipulation."[179] But by transferring the minimal definition for the characterization of mathematical work onto a machine so as to ensure that every step that it would execute could also be executed by a mathematician, Turing's machine design was elevated into an episteme. It now provides the basis for judging in the first place what a mathematician and mathematics is capable of deciding. One could not have done away more radically with unquestioned operations and habitual sign games than with this intertwining of common references to machine fictions and semiotic games into a single construction: a theoretical machine that independently inscribes or overwrites a potentially endless tape divided into squares with a finite number of different signs according to a finite number of rules. Precisely because the operation with signs is elevated here into a mathematical object of investigation, Husserl's critical question loses its validity. This is the question of the forgotten beginnings of a science, the concepts of which seem to follow a machine design while any meaningful derivation of their origins remains unaccomplished. And if Turing's proof of calculable numbers also performs the self-referential operation whereby the sign operations of his machine simulate the sign operations of another one, and is ultimately able to simulate a universal machine of all these machines defined by Turing, then Husserl's question of the "regressus ad infinitum" of a science subject to a machine design receives a negative answer, for ultimately the conclusion to which Turing comes is not positive. Turing rejected the decision problem (*Entscheidungsproblem*) that Hilbert gave up hope of solving—that every mathematical proposition of a formal system endowed with the power of arithmetic must prove through a procedure to be true or false. He did so by demonstrating the fundamental undecidability of the problem. Though a class of calculable numbers—that is, those that can be generated in an effective fashion—can be indicated, whether a real number in general is a calculable number escapes all calculability.

Turing's text reveals still more clearly than Kurt Gödel's work on formally undecidable propositions or Alonzo Church's introduction of lambda calculus that mathematical practice encounters a limit in the medium of a fictive machine. Turing's machine is a medium that does not itself possess ideal mathematical objectivity, but projects into the life-world. For that reason alone, the medium allows nonmathematical authorities in circumscribed contests with tool-like writing apparatuses their "mental manipulation" with "an infinite and yet self-enclosed world of ideal objects as a field for study." Only a mathematics that is measured against this medium reveals that it is already condemned to performance for intrinsic reasons.

Turing's discovery that if something is calculable, it can also be calculated by a machine, releases the subject from a cultural technique with a long history. But Turing's text did not simply legitimize leaving calculation to machines, for the quite simple reason that there had been a long tradition of presenting such legitimations. The fact that his fictive machine would ultimately be realized in actuality as a real discrete sign-processing machine is not so much due to the possibility of carrying out calculations efficiently. What is much more decisive is that the making-calculable—and not primarily the calculation—is performed operationally and bound to discrete signs. With the Turing machine, a machine concept appears that is based on the strictest conceivable determinism in order to reveal by that very means an entirely new form of incalculability.[180] Because sign-processing procedures potentially prove only through their execution that they come to an end, it is not enough only to imagine the Turing machine—one must let it run.[181]

It should not be overlooked, however, that the Turing machine had long asserted itself as a fictive object during the Second World War. Logician Alonzo Church, in any case, who gave Turing's fictive machine its name,[182] in the course of a career that lasted from 1924 to 1995,[183] engaged only once with automata theory, whereas computer science cannot manage without his lambda calculus.[184] But the caesura thus occurs all the more violently in the Second World War when paper machines, of all things, unleash research offensives and matériel battles.

"Merely factual sciences make merely factual people," Husserl still wrote in his *Crisis*.[185] Heinrich Scholz's noteworthy scientific career, which was defined by a shift from theology to research into the foundations of logic in the first half of the twentieth century, is able to confirm Husserl's assessment more than almost any other—but one should not therefore deny Scholz the consciousness of his own conversion. The protestant theologian was the first to respond to Oswald Spengler's radical epochal conception, and it did not elude him that Spengler—despite all the prognosticated declines—had turned above all to the coming generation: "If, under the sway of this book, people of the new generation turn to technology instead of lyric poetry, the navy instead of painting, politics instead of critique of knowledge, then they will be doing what I wish. That is the meaning of the phrase 'decline of the West.'"[186] Despite Scholz's criticism of Spengler's book, one must assume that Scholz had a Damascus experience shortly thereafter, when Whitehead and Russel's *Principia Mathematica* fell into his hands.[187] He gave up his professorship in the philosophy of religion so as to establish the first chair of mathematical logic and foundational research in Münster.

Scholz joined a circle of mathematicians that had committed itself entirely to David Hilbert's program of elucidating mathematical foundations. At the same time, Scholz was among the few who immediately recognized the significance of the fundamental work "On Computable Numbers" by the British mathematician Alan Turing. And he was the only one who asked Turing for an offprint.[188] If the Second World War had not intruded, Scholz would have ensured as early as in 1939 that Turing's fundamental insights became common mathematical knowledge through an entry in the venerable *Encyklopädie der mathematischen Wissenschaften*.[189]

Unlike Turing, who had already offered his mathematical skills to the British secret service for the investigation of cryptological methods before the outbreak of the Second World War, Scholz did not at first allow himself to be taken away from his studies of logistics, which he traced back from Gottlob Frege's *On Concept and Object* to Leibniz.[190] But in 1944 he set off for Berlin at the invitation of the still completely unknown engineer Konrad Zuse to inspect his electromechanical Z4 computing machine. An air raid siren and the resulting forced stay together in a bomb shelter caused the exchange of ideas with Konrad Zuse to go on longer than planned.[191] The connection between Zuse's concrete electromechanical computing machine and Turing's paper machine did not present itself automatically.[192] In Scholz, Zuse nonetheless found an early and prominent advocate. The arms industry, however, had no use for his machine. Meanwhile, in Bletchley Park—Great Britain's secret service establishment—Turing had managed to decrypt the German navy codes with computing machines christened "bombes," which contributed decisively to the turn of the naval war to the Allies' advantage.

Turing and his colleague Gordon Welchman had succeeded in reading self-reciprocal functions from the wiring of the German encryption and decryption machine for radio messages, the ENIGMA—a possibility that had not been foreseen on the German side. The encryption potential of the machine was thereby compromised. To avoid that would have been the task of Gisbert Hasenjäger, who had received his mathematical promotion from Scholz when he was just a high-school graduate, before he was drafted into the military and ultimately, with Scholz's mediation, ended up in the department known as Referat IVa of the OKW code section, where it was his job to ensure and enhance the cryptological efficiency of Enigma.[193] But Hasenjäger and Scholz knew nothing of Turing's cryptological operation on the British side.[194] Not until two decades after Scholz's death did the British secret service captain Frederick William Winterbotham, after long hesitation, agree to report on the undertakings in Bletchley Park.[195]

While their countries' armies of millions were worn down in the Second World War, the rather small group of foundational mathematicians and logicians had been drawn into a game-like discursive formation that could not itself achieve a complete overview of itself. Thus the strategy, cultivated for quite a while in mathematical discourse, of reducing problems to game constellations, seems now to have encompassed those mathematicians themselves and assigned them each a limited function. The reductionism that characterizes the game concept thereby stands in a strangely reciprocal relation to an ever more unrestricted concept of calculability. An event in the spring of 1945 also makes this clear, when Zuse, in the composition of his plan calculus—which would later turn out to be the first design of a programming language—moves away from the strict concept of calculability and seeks, most likely still under the impact of the war, to grasp this concept in its whole scope in everyday usage. In a programming language, the concept of calculability must not remain limited to an arithmetical level. Rather, the word *rechnen* (reckoning, computing, calculating) must encompass the whole range of meaning that it has in German, and thus also the meaning in the following example: "I reckon [*Ich rechne damit*] that the enemy will withdraw when his supply line is cut and a break-through can be successfully averted."[196] To exploit the possibilities of computing machines that go beyond the domain of the merely arithmetical, nothing seemed more appropriate than elevating the programming of games, and in particular that of chess, to a test case. Whenever military duty prevented Zuse from realizing electromechanical components as mathematical logics, he turned to the travel chess set as a substitute for developing logical calculus on a material basis.[197] When, in April 1945, not even a chessboard was available any longer, and Zuse kept his Z4 computer hidden from the advancing American troops in a hay shed in the Allgäu, he formulated the basic schema of today's standard programming languages. His "plan calculus," as he called the schema, was able to take up the game of chess along with its rules as "pure desk work." Thus, it could deal with the framework of the game and its execution within the same "two-dimensional notation" on one and the same paper.[198] The fact that Alan Turing—and on the American side the communications scientist Claude Shannon—just as passionately programmed or built the first digital computers, which were capable of mastering chess and other games,[199] without having access to knowledge of Zuse's hidden activity, awakens the suspicion that the affinity to the game arises from the medium of the computer itself. At the least, the game, as an application of the computer, illuminates the latter's potential as a platform of the highest concretion and exemplary openness. Thus Shannon, on the occasion of receiving the highest distinction for engineers in the United States, claimed of his "game playing machines" that they could master all the other challenges that lay ahead in the time after the war.[200] According to Shannon, programs that enabled a "general-purpose computer" to play chess also make it possible in the long run to translate languages, make strategic decisions in simple military operations, orchestrate a melody, or carry out logical deductions.[201]

But it was someone else who realized the first step in taking the game seriously: John von Neumann. He grants the game a value that is markedly different from its common status during the foundational debate. If there the game helped formalism get beyond tautologies by way of signs, then knowledge of the game itself was never the central interest—not even when concrete games like chess were dealt with.[202] The great reductionist Neumann set to work in a quite different direction: he homed in on the game itself and could view himself with some justification as the founder of game theory. Just as Neumann would much later advise Norbert Wiener not to study the essence of life in the complexity of measured brain waves, but rather in its simplest conceivable form, the cell, the young Neumann turned away from chess and toward the most childishly easy games: the "even and odd" game or "rock, scissors, paper." Neumann conceded that most games had more complicated rules and were, in addition, dominated by chance. Here one should consider card and dice games. Moves that are dependent on chance and the calculated steps of the players appear to demand a differentiation. Furthermore, one should assume that the interplay between the players requires an independent mathematical model: "[In particular] the consequences of the circumstance (so characteristic of all social events!) that every player has an influence on the results of all the others and at the same time is only interested in his own [must be taken into account]."[203] Neumann introduces all these differentiations and variables only to discount them. The conscious steps or chance moves are determined by the rules of the game and delineated through random distributions. In Neumann's theory of the game, there is no reason not to fix the strategy according to which one plays already before the game.[204] What can be calculated can also be calculated into a "method of play"[205] before the game. What cannot be calculated before the game also cannot be calculated in the game, and Neumann launches into an intricate proof that demonstrates the validity of his minimax theorem by means of a bilinear function. According to the theorem, methods of play exist that guarantee the highest possible wins, even if optimal methods of play are opposed to them. Only the assumption of an additional participant in a game for which Neumann succeeded in demonstrating the proof of the minimax theorem poses problems for which he can only begin to present a possible solution. His theory is unable to show whether and how minimax theorems can be calculated for the rules of random games—not even when he develops it into "game theory" in 1944 with the active participation of Oskar Morgenstern. In other words, selected games may have become calculable through Neumann; however, games as they are generally found have not.

Nonetheless, from an epistemic perspective, Neumann's "Theory of Parlor Games" is highly significant, even if he himself first comes back to his early work in the priority debate with France's great mathematician and sometime minister of the navy Émile Borel.[206]

Until Neumann developed his theory of the parlor game, he was a child prodigy who rather overtaxed his Berlin mathematics professor with a dissertation on the axiomatization of set theory.[207] A closer exchange with Hilbert's Göttingen school therefore seemed desirable. Scarcely three weeks after he had arrived in Göttingen as a Hungarian doctoral candidate on a Rockefeller fellowship,[208] Neumann gave the first public talk of his life—before the renowned Göttingen Mathematical Society. The title of the lecture was "On the Theory of Parlor Games."[209] A reputation had preceded him for having advanced the axiomatization of Zermelo-Fraenkel set theory in a fashion that made the problematic axiom of choice dispensable. If Neumann had come to Göttingen as a young fellowship holder who introduced himself with his game theory, he ultimately left as a close colleague of Hilbert's, who had made his so-called Hilbert spaces productive for the mathematical grasp of quantum mechanics. It is therefore plausible to suspect a not inconsequential strategy behind the fact that between these phases Neumann chose a "Theory of Parlor Games" for his first public appearance in Göttingen. Nonetheless, Philip Mirowski seems to be the only one thus far to attempt to unpack the epistemic content of the early game theory in the context of his other works without letting himself be led astray into mere speculations by the shimmering concepts of game theory.[210]

For Neumann's later companion and friend Stanislaw Ulam, the game is the medium that led to the theory of the computer, for Hilbert's program of consistent axiomatization followed "the goal of treating mathematics as a finite game. Here one can divine the germ of Neumann's future interest in computing machines and the 'mechanization' of proofs."[211] To be precise, it should be noted that Neumann did not immediately transition from formalist mathematics as a game with intrinsically meaningless signs to the mathematical theory of computing machines. Rather, he extracted from very concrete games mathematical problem situations that were subsumed neither in a pure formalism nor in a pure application. The fact that Neumann, after his lecture in Göttingen, devotes himself to current questions of physics and above all quantum mechanics is thus also rooted in his game theory. It made it possible first to pose questions of indeterminacies and the interplay of complex systems, and then to bring them to bear on quantum physics. With game theory as a consistent implementation of the axiomatization demanded by Hilbert—and indeed, for the first time, beyond the mathematical branches—a world picture also first appears that already reveals on a global scale what will first be proper to quantum mechanics on the smallest level according to the Copenhagen model: "That which is dependent on chance ('random,' 'statistical') is rooted so deeply in the nature of the game (if not in the nature of the world) that it is not even necessary to introduce it artificially through the game rules: even if in the formal game rules there is no trace of it, it establishes itself on its own."[212] Neumann thus sought to grasp the game theoretically at the moment when a theory of what mathematics is appeared most urgent and provided him the means for the formulation of his game theory. Neither an irrevocable determination of mathematics nor of games emerged from that. For that reason alone, Neumann's "Theory of Parlor Games" does not offer an early mathematical model of social systems. His game theory did not bring to light a kernel of social and/or economic conduct, but a method that limited itself to sign operations and that, during the Second World War, brought into knowing and unknowing contact a small but influential circle of actors in nothing but zero-sum games. Shannon, in any case, designated as "zero-sum games" that level of the Second World War on which the warring parties attempted to crack the communications of the enemy and encrypt their own.[213]

If mathematics itself nonetheless ultimately failed to produce an all-encompassing theory of the calculability of games reduced to sign operations, then it cannot be the goal of this study to present another attempt to define the game. It should, however, have become clear how games in particular, under high-tech conditions, served an ever-expanding tendency toward making-calculable as a conduit medium—and still do so today. Furthermore, it should have become clear what incalculabilities, from a historical perspective, were sometimes hazarded for that purpose.

Games, like mathematics itself, seem to be defined by a tautology that either includes or excludes everything. This unity in multiplicity is also evident in the fact that in German one speaks of *die Mathematik* in the singular, as if there were only the one, while in the English- or French-speaking world the *pluralis tantum* "mathematics" and "les mathématiques" invoke several modes of being.[214]

Still more significant is the ontic dimension that the game shares with language: Heidegger knew all too well that expressions such as *Der Raum räumt* ("space spaces") or *Die Zeit zeitigt* ("time times") subvert customary linguistic usage.[215] In German, there are at least two exceptions: *Spiele spielen* and *Sprachen sprechen* mean *to play games* and *to speak languages*. In both cases, there is no way to phrase it without the tautology of noun and verb. This idiosyncrasy of German may serve as an illustration of a more general principle: games must be played and languages must be spoken in order to be what they are. A history of games—and a history of war games in particular—here reaches its limits.

Notes

Unless otherwise indicated, all quotations from German-language sources are translated by Ross Benjamin.

Preface

1. As early as in the 1950s, Vera Riley and John P. Young came to this conclusion, exploiting the potential of the war game for the then-new scholarly discipline of operations research. Riley and Young trace the relevance of the war game back to two essential factors: first, the success it afforded the German military, which probably made "the greatest use of the war game, in the past half century" (1957, 8). Second, they cite John von Neumann's game theory, which had already been presented in 1928 and which after the Second World War made possible a far-reaching theorization of tactical and strategic games (9).

2. Clausewitz 1976, 119.

3. See, for example, Booß-Bavnbek and Høyrup 2003 or Mehrtens 1996, 87–134.

4. This is how Bernays describes the fundamental suspicion toward the specialized field of logic (1976, 1).

Chapter 1

1. See Ries 1522.

2. For a persuasive historical study of zero and nothing, see Rotman 1987.

3. Through the systematic investigation of medieval sources, medievalist Arno Borst managed to rescue from oblivion the history of the Battle of Numbers and make the *Rithmomachia* legible beyond a small circle of specialists. Borst's authoritative publications on the Battle of Numbers are *Das mittelalterliche Zahlenkampfspiel* (1986) and "Rithmimachie und Musiktheorie" (1990, 253–288).

4. Borst summarizes the genesis of the word and its various spellings. See Borst 1990, 256, 261, 281.

5. Borst 1990, 258.

6. Not until the early Scholastic period do commentators begin to designate the Battle of Numbers as a game. See Borst 1990, 256, 285.

7. The Battle of Numbers was first called by name in 1070 at the cathedral school in Lüttich (Borst 1990, 276).

8. Duke August II of Braunschweig-Lüneburg, who will be discussed further in chapter 2, took up instructions for the *Rithmomachia* as a curiosity in his famous chess book. See Selenus 1978.

9. See Borst 1986, 96.

10. Gottfried Friedlein was most likely the first to have alluded to this situation. See Friedlein 1863, 298.

11. Werner Bergmann points out that the abacus takes on the calculation of geometric figures and thereby moves away from arithmetic, which stresses numerical concepts and relations. See Bergmann 1985, 117.

12. See Borst 1986, 55.

13. See Friedlein 1863, 327, and Borst 1986, 473.

14. See Borst 1986, 278.

15. Borst on the new, stratifying format of the composite manuscript: "[the] intellectuals concerned with the Battle of Numbers were obsessed with commentary: they could scarcely hear or read something without immediately writing about it. The genre responsible for such short written preliminary notes was the composite manuscript" (1986, 276).

16. Borst 1986, 326.

17. Borst 1986, 277.

18. See Busch 1998, 126.

19. See Knobloch 1989, 243.

20. See Bischoff 1967, 256, 259.

21. Bischoff 1967, 255–259.

22. See Friedlein 1863, 313.

23. Translation based on Friedlein 1863, 299.

24. See Borst 1986 for Greek numerals in the Battle of Numbers: 132–134, 147, and for Arabic numerals, 117.

25. See Bergmann 1985, 210.

26. Bergmann 1985, 209.

27. See Vossen 1962, 139. See also Bergmann 1985, 196–197.

28. See Vossen 1962, 141–142, though the question of whether the reference is to gobar digits or Greek letter-numbers remains open. See also Bergmann 1985, 210.

29. Borst speaks out against Walther as the founder of the Battle of Numbers. Compare Borst 1986, 42–43. See also Vossen 1962, 145. Moritz Cantor, on the other hand, presumes Walther to be the founder of the game (Cantor 1922, 1:851–852).

30. Walther von Speyer, quoted in Vossen 1962, 52.

31. Illmer et al. 1987, 48–57.

32. See Borst 1986, 69.

33. Huffmann 1993, 419.

34. Domaszewski asserts, "The formulaic character of these phrases makes them recognizable as technical expressions, which are taken from the command language of the Roman military" (1885, 5–6).

35. Erdmann 1935, 30.

36. Voltmer 1988, 188–189.

37. Borst 1986, 72.

38. Borst 1986, 86.

39. Upton-Ward 1992, 89, Rule 317.

40. Fleckenstein 1980, 19–20.

Chapter 2

1. See, for example, Furttenbach 1663.

2. See Knobloch 1973–1976, 56, and Krämer 1988, 107.

3. Thus asserts Leibniz, in old age, looking back at his project in a letter to Pierre Rémond de Montmort. See Leibniz 1887, 3:605. See also Krämer 1988, 104.

4. See Alberti 1973, 3:130–173.

5. See Henniger-Voss 2002, 371–397.

6. See Edgerton 1980, 195.

7. See Mahoney 1985, 195, 210–217.

8. See the introductory chapter, "The Battle of the Scholars," in Ore 1956, which deals with the complex web of oral and printed information flows within mathematical practice in the plagiarism conflict between Girolamo Cardano and Niccolò Tartaglia. On Tartaglia's new discursive role, see Henniger-Voss 2002, 380.

9. One of the first books of this sort, which speaks of games made up of mathematical problems and means optical measuring systems, is Alberti 1973, 130–173.

10. A particular genre of books is devoted to instruments extracted from books on measuring, drawing, and tabulating that can in turn be applied to them for the making of books. See Faulhaber 1610, 1644; Bramer 1630, 1648. Faulhaber included in his work copperplates that, cut out and affixed to wooden or metal disks, become templates for scales that make it possible to determine distances without calculation. Ivo Schneider (1993, 161–164) writes on Faulhaber's drawing instruments.

11. See Furttenbach 1663, copperplate 28. The degree to which Leibniz was fascinated by the stage techniques of his time has been explored by Siegert (2003, 161–162).

12. See Klein 1936, 127–128.

13. See Schnelle 1962, 15, and Becker 1973, 191–192, in particular note 2. Becker points out that a mathematical designation is "originally a mere marking of a certain spot on the figure ('where there is an A'), later it becomes linguistically identified with the mathematical object itself. This tendency ultimately

culminates in the modern formalist mathematics: 'Where concepts are missing, a *sign* appears in due time.' (Bernays)" (Becker 1973, 191–192). See also Gow 1884, 105, 169. Gow cites Aristotle as the first to emphasize explicitly the advantages of general designations—even if only for the comparison of unknown quantities.

14. Leibniz' "Ausführliche Aufzeichnung für den Vortrag bei Kaiser Leopold I" translated from the linguistically modernized version by Bredekamp 2004, 86.

15. Thus asserts Gottfried Wilhelm Leibniz in a letter to Oldenburg of December 28, 1675. Translation based on Schnelle 1962, 16.

16. Busche 1997, XIII.

17. Available at Berlin-Brandenburgische Akademie der Wissenschaften, <http://www.leibniz-edition.de/Hilfsmittel>.

18. See Schnelle 1962, 15.

19. See Schnelle 1962, 15.

20. Leibniz 1880, 4:15–104.

21. Harsdörffer 1990.

22. Leibniz to Rémond de Montmort, January 17, 1716 (Leibniz 1887, 3:667).

23. At the beginning of the seventeenth century, citizens of Nuremberg pressed for a school reform "so that school would not be a carnificina but vere ludus." Quoted in Radbruch 1997, 21. See also Leibniz's teacher Erhard Weigel (2004, 3:7).

24. Leibniz 1880, 73.

25. The description of du Praissac's approach to military problems is from the English edition of his work "The Art of Warre or Militarie Discourses." See du Praissac 1638.

26. See Jähns 1890–1891. Jähns quotes van Haaren: "Count Wilhelm Ludwig was the first after the age of the Romans to study tactics and use his knowledge in practice. [Everard van] Reyd translated everything related to warfare from the Greek and Roman authors, and the Count then studied it in conjunction with Colonel Cornput. This happened at a large table on which all evolutions were imitated and investigated as much as possible with lead figures. I have myself seen figures of that kind" (2:881).

27. Hahlweg 1973, 610, attachment 13.

28. Harsdörffer 1990, 516.

29. Schottelius 1991.

30. See Moll 1982, 68–69.

31. See Keynes 1921, 3.

32. See Scholz 1931, 19–20.

33. Harsdörffer (1990, 412) follows his source, as the comparison with the edition of Sieur du Praissac (1639, 1–8, supplement to the main text) showed.

34. See du Praissac: "These common places may be applied as well to divers other actions as to that of warre, provided that you know which to choose, and how many" (1639, 8).

35. See Voisé 1967, 196–197.

36. Selenus 1978, 4.

37. Selenus 1978, 111.
38. Weickmann 1664, 5.
39. Faulhaber's proportional dividers, which themselves were to be extracted from books and cut out of paper, could have been helpful in this.
40. Weickmann 1664, 7.
41. Weickmann 1664, 36.
42. Marion Faber assumes the portrait of Leopold I in this potentate. See Faber 1988, 100.
43. Faber 1988, 167.
44. Faber 1988, 168.
45. Faber 1988, 169.
46. Faber 1988, 7.
47. Leibniz 1840, 2:493.
48. Leibniz 1986a, 3:901.
49. Leibniz 1986b, 3:589.
50. Leibniz 1986b, 585.
51. Leibniz points out that "the old Germans were (so to speak) born soldiers, today they must be made into them through art and diligence" (1986b, 578).
52. See Struve and Struve 1997, 112–122.
53. See de Mora Charles 1992, 125–158.
54. Harnack 1900, 1:91. Leibniz's proposal was realized under the reign of Friedrich II and entrusted Leonard Euler at the academy with the calculation of the expected winnings in the lottery. See Maistrov 1974, 101–103.
55. In the letter to Rémond de Montmort already mentioned in footnote 22, Leibniz revisits this proposition and refines it to the effect that people are not at their most inventive in games, but rather in the invention of them.
56. Leibniz 1710, 23.
57. Leibniz 1710, 26. English translation by Richard J. Pulskamp, Department of Mathematics & Computer Science, Xavier University, Cincinnati, OH. Retrieved December 9, 2009; available at <http://www.cs.xu.edu/math/Sources/Leibniz/sinica -latin-english.pdf>.
58. Heidegger 1997, 151.
59. Heidegger 1997, 167.

Chapter 3

1. See Poten 1889, 3:153.
2. See Reiswitz 1812, XI.
3. Anonymous 1869, 276. According to Konstantin von Altrock, hidden behind the author abbreviation 98 is Theodor von Troschke.
4. It should nonetheless be noted that Kleist was also keenly attuned to the news of more recent media: he disseminated the announcement of Sömmerling's electrolytic telegraph in the form of the daily newspaper he had developed.

5. See Kittler 1988, 56–68.

6. Friedrich Christoph Dahlmann quoted in Sembdner 1957a, 233, Document No. 316.

7. Dahlmann quoted in Sembdner 1957a, 233, Document No. 317.

8. Dahlmann quoted in Sembdner, 1957a, 235, Document No. 318a. Almost half a century later, the memory of the military games remains present for Dahlmann and Pfuel, as a meeting of the two of them documents: "'Dahlmann!—40 years! Remember Prague, Kleist, the military game?' 'Indeed, Your Excellency,' I replied, 'it is 47 years, I am a historian.'" Friedrich Christoph Dahlmann quoted in Sembdner 1957a, 235, Document No. 318b.

9. Sembdner 1957a, 233.

10. Scharnhorst 1973, 49.

11. See Robb 1994, 189.

12. de Balzac 1981, 12:653.

13. de Balzac 1967, 1:27–28. Honoré de Balzac's draft is dated January 1833.

14. Reiswitz 1812, XXVI.

15. This study is preserved in the German federal military archive (Hofmann 1951). See also Hausrath 1971, 28.

16. Hofmann 1951, 28–29.

17. The half-official biographical reference book reports informatively on this situation; the omission of the name of a national writer is no less eloquent (Priesdorff 1937, 7:346–347).

18. Reiche 1857, 166–167.

19. Clausewitz 1980, 1076.

20. Dahlmann, quoted in Sembdner 1957a, 232, Document No. 317.

21. See Priesdorff 1937, 347.

22. Marie von Kleist to Frederick William III of Prussia, December 26, 1811, in Sembdner 1957a, 343. Document No. 509a.

23. Marie von Kleist to Frederick William III, September 9, 1811, in Sembdner 1957a, 339–340, Document No. 507a.

24. See Kittler 1988, 56–68.

25. Sembdner 1957b, 166.

26. See Paret 1982, 130–139. In the course of his comparison of two careers, Paret mentions an actual encounter between Clausewitz and Kleist. Indirectly, women seem once again to have sought out contacts and disseminated information quite deliberately. Marie von Kleist was a close friend of Clausewitz's later wife Marie von Brühl. It was her brother Karl who ultimately brought Kleist's play "The Prince of Homburg" onto the stage. And it is another friend of Marie von Kleist's, Frau von der Marwitz, born Countess Moltke, who explains to Kleist regarding the king: "He will never change his nature; eternally irresolute, he will thwart all well-calculated plans and cripple the strength of those who would sacrifice themselves for him." Quoted in Sembdner 1957b, 166.

27. Ludendorff, quoted in Miksche 1976, 102.

28. Clausewitz 1966, 1:234–235.
29. See Cochenhausen 1936a, 6. Van Creveld dates the breakthrough of mission-type tactics to Moltke's era (1982, 35–36).
30. Clausewitz 2007, 64.
31. Kant 1996, 18n. Emphasis in original.
32. Kant 1996, 8. Emphasis in original.
33. Kant 1996, 7.
34. Kant 1996, 7.
35. Kant 1996, 10.
36. Kant 1996, 10.
37. In this connection, Kleist's so-called "Kant crisis" is revealing. Kleist famously pointed out after reading Kant that green glasses instead of eyes would obscure the fact that not all things are green, and correspondingly eyes never show what they obscure eternally about things. The same applies to the understanding. Compare Heinrich von Kleist to Wilhelmine von Zenge, March 22, 1801, Zenge, Wilhelimine von, in Kleist 1961, 634. Ernst Cassirer sought to dismiss Kleist's reading of Kant as a misunderstanding. Compare Cassirer 1921, 170–171. Cassirer counts Kleist's example among the judgments of taste, arguing that the understanding of colors was explicitly excluded from Kant's transcendental philosophy. Nonetheless, Cassirer's objection misses the core of the matter and is tantamount to a criticism that, for example, in the four-color problem sees the problem in the selection of the colors, not in their distribution on a map. Kleist's alarming reading of Kant is, however, an early testimony to the awakening consciousness that revealing and concealing "natural phenomena" are no longer separable from processes of perception and understanding. What was adumbrated by Kleist would be spelled out later in a famous press briefing by U.S. Secretary of Defense Donald Rumsfeld to create a rationale for the second Iraq war: "As we know, there are known knowns. There are things we know we know. We also know there are known unknowns. That is to say, we know there are some things we do not know. But there are also unknown unknowns, the ones we don't know we don't know."
38. See Bosse 1990, 33.
39. See Hossbach 1954, 142–143.
40. Clausewitz to Gneisenau, June 24, 1810, in Clausewitz 1966, 1:682. See also Paret 1985, 145.
41. Clausewitz to Gneisenau, June 17 1811, Clausewitz 1966, 644. See Paret 1985, 187.
42. Paret 1985, 189.
43. Clausewitz 1966, 237–239. See Paret 1985, 189.
44. Clausewitz 1980, 1080.
45. At least formally, Wilhelm would, at the end of his reign, have the supreme power of command in the German Empire.
46. Paret 1985, 194.

47. Reiche 1857, 220.

48. Anonymous 1874, 693.

49. The text against which Clausewitz polemicizes: Bülow 1799.

50. Bülow 1799, 693, and Reiswitz's friend, General of the Infantry Ernst Heinrich Dannhauer (1874, 531).

51. Anonymous 1874, 693.

52. Anonymous 1874, 698.

53. Anonymous 1874, 694.

54. Anonymous 1874, 694.

55. For Reiswitz's predecessors and for the historiographical subject of war games in general, see Hohrath 2000, 145.

56. Hohrath 2000, 146–147.

57. That is what Reiswitz states in the German title of his manual.

58. Clausewitz 2007, 26–27.

59. See Beyerchen 1992, 59–90.

60. Clausewitz 1980, 1083.

61. In the earliest form of the tactical war game, the time interval was one minute.

62. Reiswitz 1812, 5.

63. Reiswitz 1812, IX.

64. Reiswitz 1816, VI.

65. Dannhauer 1874, 529.

66. Müffling 1824, 2973.

67. Reiswitz 1812, 3.

68. See Dannhauer 1874, 528, and Reiswitz 1824, 9–10.

69. See Kittler 1991, 134–147.

70. Grüger and Schnadt 2000, 116.

71. Altrock 1908, 165.

72. Dannhauer 1874, 530.

73. Lüdecke 2002, 30.

74. Moltke 1911, 466.

75. Moltke 1911, 466.

76. See Brandes 1902, 1:24.

77. Along with Reiswitz's friends, the later General Griesheim and Dannhauer, the latter names alongside Moltke, among others, Finckenstein and Verdy du Vernois. See Dannhauer 1874, 531.

78. Altrock 1908, 152–159.

79. Anonymous 1828, 78.

Chapter 4

1. Still, for Alexandre Kojève, the leading figure of the French antihermeneutic philosophers, Plettenberg developed an appeal that made any other invitation, even

that of the Free University of Berlin, seem uninteresting in comparison. See Taubes 1987, 24.

2. This is the famous and oft-cited title with which Schmitt's former pupil Waldemar Gurian labeled him after his apologia for the "Röhm-Putsch." See Paul Müller (alias Waldemar Gurian) 1943, 567.

3. See Schmitt's foreword to *Hamlet. Sohn der Maria Stuart* by Lilian Winstanley (1952, 7–25). Winstanley's book was originally published as *Hamlet and the Scottish Succession: Being an Examination of the Relations of the Play of Hamlet to the Scottish Succession and the Essex Conspiracy* (1921) and was translated by Carl Schmitt's daughter Anima Schmitt. See also Schmitt 1956, 73.

4. Walter Benjamin to Schmitt, December 9, 1930, in Benjamin 1997, 3:558. For an English translation of the letter, see Weber 1992, 5.

5. Schmitt 1956, 66.

6. Schmitt 1956, 42.

7. See Schmitt 1956, 62, 66–67. In terms of constitutional history, Schmitt argues in another excursus that the adherence to the medieval-feudal, sacral right of blood condemned the Stuarts to hopelessness and caused their downfall when a new elective monarchy was established. Here it is important to note, however, that "elective" by no means indicates a democratic form of state, but rather the limitation of the divine right of the king. See Schmitt 1956, Exkurs I, 57–61.

8. Schmitt even goes so far as to claim that Hamlet's hesitation is a manifestation of Shakespeare's attitude of avoiding taking sides lest he put his own neck on the line (Schmitt 1956, 20–21).

9. Schmitt, "Vorwort," in Winstanley 1952, 13.

10. See Schmitt 1956, 43.

11. Schmitt, "Vorwort," in Winstanley 1952, 12.

12. A primal scene for Carl Schmitt, incidentally: the day that the Bavarian Soviet republic was proclaimed, as assessor of the city commandant's office in Munich he had "gone to work as usual, somewhat later Schmitt and his colleagues were interrupted by revolutionaries and one of them shot an officer next to the desk" (Kennedy 1988, 147).

13. This was Schmitt's own understanding of his role, quoted in Pyta and Seiberth 1999, 424.

14. See Schmitt 1934, 945–950.

15. See, for instance, the article by Anonymous 1936, 3.

16. Schmitt's equation of himself with Herman Melville's character of the ship captain Benito Cereno, which he makes in his letters to Ernst Jünger in 1941, points in the same direction. In Melville's novel of the same name, Benito Cereno, as the white captain of a black crew that has mutinied, is forced to keep up the appearance of the sovereign commander. See Carl Schmitt to Ernst Jünger, September 17, 1941, in Schmitt 1999, 128–130.

17. Manstein 1958, 131.

18. Manstein 1958, 133.
19. Manstein 1958, 107. Gaines Post points out that the cooperation between the foreign office and the troop office was initiated by Blomberg. See Post 1973, 209.
20. Manstein 1958, 106.
21. Manstein 1958, 131.
22. Manstein 1958, 131.
23. The ministry, established in 1929, subsumed the adjutancy directly subordinate to the Reichswehr minister, the budget sections of the army and navy, the legal section, the Wehrmacht section, and the intelligence section (known as the Abwehr).
24. See Eugen Ott's retrospective account (1965). The facts are confirmed by Huber 1988, 40.
25. Ott 1965, 7.
26. See Carsten 1966, 296–308.
27. Manstein 1958, 132–133.
28. Manstein 1958, 133.
29. Spieß and Lichtenstein 1979, 27. The uniforms were provided at Hitler's behest by the foreign intelligence (Ausland/Abwehr) department of the Wehrmacht supreme command. See Spieß and Lichtenstein 1979, 38. Incidentally, Manstein—who had himself sought to obtain Polish uniforms—thus had to cede the field to Himmler and Heydrich. Erwin Lahousen Edler von Vivremont noted in the war diary of Abwehr II: "In response to my inquiry as to why General Manstein's request regarding the deployment of 3 Sturm battalions with Polish uniforms was rejected, but in the same area an undertaking of SS-Reichsführer Himmler shall be carried out, the answer is that this is the desire of the Führer, who wants to keep the Wehrmacht out of all undertakings that have a manifestly illegal character" (quoted in Spieß and Lichtenstein 1979, 39).
30. Spieß and Lichtenstein 1979, 138–140.
31. Spieß and Lichtenstein 1979, 147.
32. See Baudrillard 1993, 55–61.
33. See Baudrillard 1994, 6.
34. Hofmann 1979, 30. See also Praun (General of the Signals Corps), Militärarchiv (1951, 203). There it is asserted: "[The] signals battalions actually [make] the communications connections required in the serious case [Ernstfall]."
35. Hofmann 1979, 30.
36. See Praun 1951, 205–206.
37. Hofmann 1979, 30–31.
38. Praun 1951, 193.
39. Fangohr 1951, 131, 133, 135.
40. Fangohr 1951, 140.
41. This is Praun's conclusion after a comparison between the failed "Officer and Communications Command Post Exercise" in Silesia in 1935 and the successful one between Kassel and Rhön in 1939. See Praun 1951, 207–209.

42. Praun 1951, 201, original emphasis.
43. Fangohr 1951, 77.
44. See Huber 1988, 61.
45. See Ott 1965, 7.
46. See Huber 1988, 46.
47. The highest judicial decisions also touch on a fictional element, as Schmitt found in Hans Vaihinger's philosophy of the "as if." According to Schmitt, the domain of fictional hypotheses should not be excluded from the consideration of the facts of a case, for they—as Wagner put it—"never manage without some delusion," as quoted in Schmitt 1911, 429–430, which see Schmitt 1913, 804–805; Schmitt 1912, 239–241. For a general discussion of the significance of fiction in Schmitt's early work, see Villinger 1992, 191–222.
48. See Huber 1988.
49. The thesis was initially advanced by Papen himself; see Papen 1952, 247–249. Carsten too took it up (1966, 378–384). Ott disputed the thesis (1965, 2), correctly, as Huber illustrates (1988, 46), as does Pyta (1992, 385–428).
50. Even Wolfram Pyta, who has conducted the most comprehensive research to date on the regime of Papen and Schleicher, is susceptible to the charge of drawing false conclusions with respect to the war game, due to the limits of the time frame under investigation and the reduction of the perspective to the political level. When he points out that after the beginning of Schleicher's chancellorship, the emergency plans that were derived from Ott's war game were "by no means shelved, but held in reserve as a political option," his formulation of the question excludes any possibility of a military-strategic connection. See Pyta 1992, 387. Pyta takes into account neither that other war games followed Ott's nor that others had preceded it; he regards Ott's war game as simply "a novelty: it was held because the Reichswehr in 1932 had a qualitatively different and more difficult task to manage than in 1923–24" (388). Yet he could have found in Ott's account itself that at least one other war game had preceded his—indeed, one with the same domestic and foreign policy orientation.
51. See also Post 1973, 320. On the basis of Ott's game, Pyta comes to the opposite conclusion—that the potential resistance of the NSDAP and KPD had actually been considerably less than assumed by the direction of the war game, and emergency decrees might well have been realizable (1992, 392). The potential threat in the case of an attack by Poland, however, he judges as even higher than assumed by the war game participants: "if it [the Reichswehr] already reckoned with Polish encroachments, to be safe it should have extended this scenario [beyond East Prussia] to Eastern Pomerania and Upper Silesia as well. By omitting this step in the war game and limiting a threat to German borders solely to East Prussia, the enacted situation was even prettified in comparison to the perceived threat actually hovering before the Reichswehr leadership" (390). In other words, the domestic situation might have been controllable through an emergency decree, but for a conflict with Poland, the presidential regime would no longer

have been equipped. Pyta's conclusion that the presidential regime had portrayed the actual danger in the war game as greater than it was and thereby squandered its political sphere of influence stands in contradiction to the facts that he himself brought to light.

52. See Ott 1965, 4–5.

53. Pyta 1998, 177.

54. Pyta 1999, 417–441.

55. See Pyta 1999, 417–441, and 1998, 178.

56. See Huber 1988, 40.

57. Michael 1999, 438.

58. Documents on Ott's war game and commentary in Pyta 1992, 385–395, 395–414. Pyta rejects the concept *Kriegsspiel* ("war game") in this context (387). Still, *Kriegsspiel* is used as an umbrella term in the military and also in this context by participants such as Ott. In the military sources, however, *Kriegsspiel* is employed not only as an umbrella term, but also in contrast to *Planspiel*, though in this case *Kriegsspiel* signifies the playing through of a war scenario by two equal parties, and in the *Planspiel* the leadership of one party takes over the direction of the game. See List 1951, 148.

59. See Pyta 1992, 400.

60. See Pyta 1999, 432.

61. See Ott 1965, 11.

62. See Pyta 1992, 403. In the documents of the war game, one can read descriptions like these: "The long-distance connection to Berlin has been destroyed since November 24 at 21:00"; "The Prussian Ministry of the Interior on Unter den Linden was badly damaged by an explosives attack. . . . In northeastern Berlin (Stettiner Bahnhof, Nordbahnhof) intense disturbances as a result of terror in Wedding"; "Acts of sabotage on railroad cars and in vital occupations must be expected in the days to come." All quotes are taken from the documents Pyta includes in "Vorbereitung für den militärischen Ausnahmezustand" (1992, 403–405).

63. See Krosigk 1989, 2:1037. See also Pyta 1992, 387, as well as "Vortragsnotiz des Oberstlt. Ott für den Reichswehrminister von Schleicher," in Vogelsang 1962, 484–485.

64. See Huber 1988, 65. See also Pyta: in "the particularly thorny constitutional matters, it was not the 'minister of the constitution' Gayl, whose province they actually were, who was responsible, but the Reichswehr minister for operational planning" (1998, 177).

65. Anonymous, quoted in Pyta 1992, 407.

66. See Pyta 1992, 393, 415n77.

67. See Huber 1988, 49.

68. See Dirks and Janßen 1999, 45. The portrayal is based, according to friendly information from Carl Dirks, on the document "Erste operative Aufgabe des Truppenamtschefs vom 10.1.1933," Nr. 899/32. (Archiv Dirks).

69. See Pyta 1998, 178, 191n55.

70. When Hitler, after the invasion of France in late July of 1940, decided to go to war against the Soviet Union, Marcks—as the chief of the General Staff of the 18th Army—had already long ago been tasked by Franz Halder, without orders from the "Führer," with designing a first operational plan against Russia. See Ueberschär 1998, 24. See also Dirks and Janßen 1999, 138, and Hofmann 1979, 67. Marcks's operational plan, which identified Moscow as a decisive military objective—a position from which Hitler famously distanced himself—is reproduced in Ueberschär 1998, 223–234. The development of the overall operation "Barbarossa" is described by Hofmann at the express request of the Historical Division of the American Army as an iterative process that constantly supersedes personal responsibilities: the first plans were developed independently by Marcks, the operations section of the army high command, the chief of the army General Staff, as well as—somewhat later—General Friedrich Paulus. Based on the plans, the operations section then prepared marching orders, which were in turn tested in war games. The general staff of the army took the same steps independently and compared its results with those of the high command. Thereafter, the "Directive for Barbarossa" was issued on December 18, 1940, by the supreme commander of the Wehrmacht. Hitler's directive then underwent a precise elaboration in Army Group A under the leadership of General Sodenstern in the form of the map exercise "Otto." Map exercises at the level of the army group high commands followed. Finally, the marching orders of the high command ensued on January 31, 1941. Again it was through map exercises that the panzer group commands and individual division commands became acquainted with the implications of the orders. See Hofmann 1979, 46–63, 67–71.

71. See Finker 1994, 239.

72. The "ingenious idea," according to Hans Mommsen, does not stem from Colonel Henning von Tresckow, but, as Dirks and Janßen claim, from Friedrich Olbricht, who already had the idea in the winter of 1941–1942. See Dirks and Janßen 1999, 170. For an opposing view, see Page 1989, 186, and Finker 1994, 238–239. The fact that Olbricht's "Valkyrie Plan" was not a stroke of genius but a new version of Ott's war game, though under a different sign, is suggested by Olbricht's career in the troop office in section T3 (foreign armies); its two sections T1 and T2 were decisively incorporated into Ott's war game at the same time. See Anonymous, "Ausgangslage der Planübung," in Pyta 1992, 396–397. When Olbricht took over the leadership of the successor institution to the troop office, the general army office, in 1940, he had access to all the files on the current emergency plans. Their reworking ensued with the familiar General Staff–like routine: "It is all too often forgotten—especially among those who seek to stress Stauffenberg's role in the plot—that the official 'Valkyrie' orders were revised a total of 220 times, and the version of July 31, 1943—still 2 months before Stauffenberg took office and even before his initiation into the plot—was already the 83rd version of these orders!" (Page 1989, 190).

73. See Hoffmann 2007.

74. Speer 1996, 463.

75. An instructive discussion of Verdy du Vernois's use of the applicatory method is provided by Lange (2003, 218–239). However, he dates the method's origins too late. On the supposed beginnings of the applicatory method, see Lange 2003, 225. The fact that the applicatory method actually coincides with mathematical training at the time of the establishment of the General War Academy in Berlin will become apparent through the analysis in the next chapter.

76. Verdy du Vernois 1881, VIII.

77. The debate was instigated by an article by Hans Meier-Welcker, "Unterricht und Studium in der Kriegsgeschichte angesichts der radikalen Wandlung im Kriegswesen" (1960, 608–612). A collection of the contributions to the debate, which then unfolded between Meier-Welcker, Hermann Heidegger, Friedrich Forstmeier, and Gerhard Papke can be found in Messerschmidt 1982.

78. Heidegger 1961, 196.

79. Heidegger 1961, 195.

80. Ernst Jünger, quoted in Heidegger 1961, 195.

81. Heidegger 1961, 197.

82. Heidegger 1961, 197.

83. Schellendorff 1875, 1:138.

84. See Brühl 1973, 80.

85. Otto 1993, 427.

86. Otto 1993, 431.

87. Otto 1993, 432.

88. See Reiswitz 1824, X.

89. Anonymous [Albrecht von Roon] 1854, 8.

90. Anonymous [Albrecht von Roon] 1854, 3.

91. Anonymous [Albrecht von Roon] 1854, 8.

92. Johann Eduard Erdmann, quoted in Nicolin 1971, 442.

93. Anonymous [Albrecht von Roon] 1854, 9.

94. Anonymous, Reichsgesetzblatt 1919, Teil I, Nr. 140, quoted in Model 1968, 21.

95. See Huber 1937, 267–268.

96. Even the power of command during the First World War, which rested, according to the constitution, with Wilhelm II, but which he did not exercise, already shows how the means and methods of invoking it are actually more decisive than its normative possession. The head of state law of 1934, which assigned the supreme command over the Reichswehr to the Führer and Reich Chancellor, did not bring about a restoration of the power relations of the Wilhelmine era because in actuality, overlapping claims to the power of command increased instead of decreasing. What other army would have permitted a Wehrmacht high command and an army high command with overlapping domains of power?

97. See Otto 1993, 423. After the dissolution of the Great General Staff on September 30, 1919, Reich President Friedrich Ebert decreed the founding of a Reich archive that was directly subordinate to the Reich Ministry of the Interior.

98. Hermann Göring, quoted in Stahl 1977, 72.

99. See Hartlaub 2002, 2:330.

100. Felix Hartlaub to Melita Laenebach, January 15, 1943 (Hartlaub 2002, 1: 567).

101. Hartlaub 2002, 1:568.

102. Hartlaub 2002, 1:568.

103. Hartlaub 2002, 1:568.

104. Hartlaub 2002, 2:24.

105. Hartlaub's immediate superior in the military history section of the Wehrmacht high command was the medievalist Percy Ernst Schramm.

106. See Marose 2005, 151–152.

107. Hartlaub, "Im Dickicht des Südostens" in Hartlaub 2002, 1:187.

108. Hartlaub 2002, 1:191–192.

109. Hartlaub 2002, 1:187.

110. Hartlaub 2002, 1:188.

111. In fact, the map table under which Stauffenberg left his briefcase prepared with explosives and time fuses protected Hitler from the main force of the explosion. Thus his blood was not on the "tattered bloody maps" that were offered to Hartlaub—as the deliverer commented, "military history" must attach "the utmost importance" to something of this kind (Hartlaub 2002, 188).

112. See Kirchmann 1998, 16.

113. Virilio 2005, 111. Emphasis in original.

114. See Virilio 2005, 111.

115. Archiv der Humboldt-Universität zu Berlin, email message to the author, January 2004.

116. See Hildebrand 1990, 161–162. The Humboldt University archive could not determine that Cochenhausen received a doctorate from the Friedrich Wilhelm University.

117. Cochenhausen 1934b.

118. Groener 1930. Among the contributing Reichswehr officers is Schmitt's close acquaintance Erich Marcks. His conception of a Führer's life states, "He stopped loving people early on; he saw through them, and the pettiness and lowliness of the sentiments that he had encountered among so many he soon assumed of all. Whoever knew him did not love him; but all soon succumbed to the conquering power that emanated from him. His figure was still gaunt, his face yellowish and angular, but the certainty of his nature was imposing, and from his gray eyes blazed the flame of genius. With consummate art he knew how to deal with people: their drives, patriotism and longing for glory, ambition and greed, love and fear were like an instrument to him, which he understood how to play at will. He completely controlled the temperaments of his soldiers.

He knew so fully how to press their noble and base passions into his service that soon the striving for military heroism took the place of 'civic virtue,' the cult of his personality took the place of the ideal of the republic" (259). He is not speaking of a contemporary dictator, but of Napoleon!

119. Cochenhausen 1934a.

120. Foucault 2003, 82.

121. Quoted in van Creveld 1982, 141.

122. A publication of the American Management Association on a business game that is regarded as the first ever reads, "In the war games conducted by the Armed Forces, command officers of the Army, Navy, and Air Force have an opportunity to practice decision making creatively in a myriad of hypothetical yet true-to-life competitive situations. Moreover, they are forced to make decisions in areas outside their own specialty; a naval communications officer, for example, may play the role of a task force commander. Why then, shouldn't businessmen have the same opportunity? Why shouldn't a vice president, say, in charge of advertising have a chance to play the role of company president for fun and for practice? Why not a business 'war game,' in which teams of executives would make basic decisions of the kind that face every top management—and would see the results immediately? From these questions grew AMA's Top Management Decision Simulation. After an exploratory visit to the Naval War College, a research group was formed and work began on a game which would eventually become part of an AMA course in decision making. This in turn, it was hoped, might lead to a sort of 'war college' for business executives" Ricciardi et al. 1957, 59. IBM and McKinsey & Company were then the first to implement the business games on computers on the one hand and in paper form on the other. See Cohen and Rhenman 1961, 131–166. For a general overview, see Hausrath 1971.

123. Cochenhausen 1936b, 1:102.

124. According to Martin van Creveld, the German Army differed fundamentally in this respect from the armed forces of the United States (1982, 144).

125. Cochenhausen 1936b, 102.

126. Cochenhausen 1936b, 103.

127. Taysen 1936, 99.

128. Voigt 1975, 231–260.

129. Voigt 1975, 250–251.

130. Voigt 1975, 250–251.

131. Voigt 1975, 249.

132. Voigt 1975, 249.

133. Voigt 1975, 259.

134. As is well known, the ranks of the SS ran from "Reichsführer" Himmler, through the "Obergruppenführer," "Oberführer," "Obersturmbannführer," "Standartenführer," and so on, down to "Rottenführer."

135. The single exception in the linguistic usage of the Reichswehr: the term "Führergehilfe" used in the training of General Staff officers.

136. Thus, for example, one spoke again in all sorts of connections of the General Staff or the Quartermaster General.
137. Finker 1994, 239.
138. Finker 1994, 239.
139. Page 1989, 190–191. Emphasis in original.
140. In the case of Stauffenberg, it is not even an older leadership technique that must supersede the new one, but the secret-society oath sworn to George's "Secret Germany" that legitimizes and demands the break of the loyalty oath to the "Führer and Reich Chancellor." The fact that George himself might have "overestimated the role of poetry [in the political]" as the poet Thomas Kling argues (with Ernst Osterkamp) did not occur to Stauffenberg. See Kling 2005, 65.
141. Schmitt 1936, 1:549.
142. Hitler, quoted in Domarus 1965, 2:1316.
143. Fangohr 1951, 83.
144. List 1951, 157.
145. List 1951, 154.
146. van Creveld 1982, 5.
147. van Creveld 1982, 28.
148. Sigmund Freud, "Massenpsychologie und Ich-Analyse," in Freud 1940, 106.
149. This discursive reorientation of psychiatry is described in detail by Wolfgang Schäffner (1991, 25–33).
150. Schäffner 1991, 44–50.
151. Anonymous, "Felddienstordnung von 1908," quoted in van Creveld 1982, 36.
152. van Creveld 1982, 36.
153. Arbib 2000, 195. See also Arbib 1972, 525. See also Stafford Beer, according to whom McCulloch traced the command technique back to Nelson (Beer 1993, 33). For the valuable reference to the concept of the "redundancy of potential command" in this context, I am indebted to Michael Mahoney. For the unexpected insight that the German ground forces—in contrast to the American—took up this command technique, I am indebted to Friedrich Kittler.
154. Arbib 1972, 203.
155. Arbib 1972, 212.
156. List 1951, 166–167.
157. In retrospect, the fictional hypothesis of the captured "black boxes" appears to refer to a specific kernel of truth, which was prohibited from being mentioned for reasons of utmost secrecy. Among all the captured enemy military apparatuses, the German encryption and decryption machine ENIGMA was assigned the greatest significance. The decryption of German radio messages was decisive for the submarine war in the Atlantic. Even though the British repeatedly managed to capture ENIGMAs or parts of them before the German crews could destroy them, the efforts ultimately amounted to deriving the last technical configuration of the ENIGMA solely from the nature of the encrypted radio

messages. Wiener and Neumann's fictive scenario thus coincides with the actual efforts in Bletchley Park in Great Britain, where above all the mathematician Alan Turing and Gordon Welchman with the help of so-called bombes—electronic special computers—advanced the decryption of German naval radio messages. The undertakings of the British cryptologists were, however, scarcely known at this time outside England. Whether Neumann, who spent time there in 1943, was in some way initiated, is questionable. But regardless of whether the invention of the fictitious "black box" anticipated or covered up a highly specialized communications technology, with the emergence of systems theory it will provide the model by which every form of communication must be judged.

158. McCulloch 1974, 11. See Arbib 2000, 200.

159. It would not be the last exchange of blows between the two of them. This time, however, they reportedly went arm in arm to lunch. See Arbib 2000, 200.

Chapter 5

1. R. G. D. Richardson (Brown University Providence) to Douglas C. Jackson (MIT), November 19, 1942, the Institute Archives and Special Collections, NWP, box 4, folder 62, MIT Libraries.

2. See Genuth 1988.

3. See Galison 1994.

4. See Galison 1994, 242.

5. Quoted in Lorey 1916, 3:31.

6. Scharlau and Knobloch 1989.

7. Müffling to Bernhard von Lindenau, April 1, 1823 (Bruhns 1877, 11).

8. Bruhns 1877, 10.

9. Humboldt to Carl Friedrich Gauß, July 12, 1849 (Bruhns 1877, 55).

10. Müffling to Gauß, April 14, 1821 (Bruhns 1877, 8).

11. Humboldt to Gustav Dirichlet, September 16, 1828 (Biermann 1982, 49).

12. Humboldt to Gauß, September 8, 1828 (Bruhns 1877, 22).

13. Müffling to Lindenau, November 28, 1824 (Bruhns 1877, 13).

14. Biermann 1958/1959, 125.

15. Humboldt to Dirichlet, April 9, 1828 (Biermann 1982, 38).

16. See Löbel 1992, 144, 146, and Lüdecke 2002, 7–8.

17. See Lampe 1906, 483.

18. Humboldt to Dirichlet, September 22, 1828 (Biermann 1982, 50).

19. See Biermann 1982, 41.

20. Pieper 2003, 28.

21. For a detailed account of this, see Siegert 2003, 240–252.

22. Kummer 1988, 52.

23. Humboldt to Dirichlet, May 27, 1828 (Biermann 1982, 43–44).

24. Humboldt to Dirichlet, June 12, 1828 (Biermann 1982, 45).

25. Hensel 1988, 91.
26. Ambroise Fourcy, quoted in Manegold 1966, 183.
27. Jacobi 1891, 7:357.
28. Jacobi 1891, 366.
29. Humboldt, quoted in Biermann 1982, 32–33.
30. Lampe 1906, 483.
31. Lampe 1906, 483.
32. Hassel 1905, 24.
33. Hassel 1905, 24.
34. Rühle von Lilienstern to Karl v. Altenstein, July 4, 1828 (Biermann 1959, 41–42).
35. Dirichlet's request to audit courses at the school was thwarted by the Prussian chargé d'affaires in Paris. See Kummer 1988, 41.
36. Scharfenort 1910, 112.
37. Humboldt to Dirichlet, February 22, 1829 (Biermann 1982, 53).
38. Well-informed about this, thanks to his position there, was Lampe (1906, 484).
39. Lampe 1906, 484.
40. See Butzer, Jansen, and Zilles 1984, 9.
41. Klein 1904, 267–276; see in particular 274–275.
42. Humboldt, quoted in Biermann 1988, 39.
43. Biermann 1959, 43.
44. Aschhoff 1995, 2:68–69, 95–98.
45. Biermann 1958/1959, 124.
46. Hohenlohe, quoted in Scharfenort 1910, 161.
47. August Leopold Crelle, quoted in Manegold 1966, 190.
48. Du Bois-Reymond 1910, 195.
49. Du Bois-Reymond 1910, 198. See also Bourbaki 1948, 46–47.
50. Wittgenstein 2009, 25.

Chapter 6

1. Wittgenstein 1991, 68.
2. Wittgenstein 1991, 68.
3. Wittgenstein 1991, 131.
4. Wittgenstein 1991, 50.
5. Wittgenstein 1991, 130.
6. Wittgenstein 1991, 47.
7. Wittgenstein 1991, 69.
8. Wittgenstein 1991, 70.
9. Wittgenstein 1991, 57.
10. Wittgenstein 1991, 70. Wittgenstein's Kafkaesque description of the position: "Am like the prince in the enchanted castle on the observation post."

11. See Zabecki 1994, 2.

12. Zabecki 1994, 7.

13. See Zabecki 1994, 17.

14. Blumenberg 1990, N3.

15. Wittgenstein 1961, 2. (The publishers did not include the cryptographically written part of Wittgenstein's diaries and concealed its existence. See Wittgenstein 1991, 159–186.) On the question of logic, see also Blumenberg 1990, N3.

16. On the right side of the diary, Wittgenstein recorded the logical reflections, and on the left, ruminations on his personal situation—encrypted with a Caesar cipher.

17. Wittgenstein 1991, 42.

18. Wittgenstein 1991, 136.

19. Wittgenstein 1991, 137.

20. Stone 1975, 252–254.

21. See Zabecki 1994, 27–28.

22. See Zabecki 1994, 27–28.

23. Bruchmüller 1922, 109.

24. Bruchmüller 1922, 109.

25. See Zabecki 1994, 29.

26. See Kittler 1996, 205–207, 219.

27. See Kittler 1996, 209; Lupfer 1981, 8–12, 42; and Zabecki 1994, 2.

28. See Kittler 1996, 207.

29. Kittler 1996, 206.

30. See Bruchmüller 1922, 29–30.

31. Bruchmüller 1922, 82.

32. Bruchmüller speaks of the unified "regulation" of a creeping barrage, "whereby the independence of the subordinate officers is admittedly constricted to a certain degree" (Bruchmüller 1922, 109).

33. That might be the reason that the so-called Pulkowski method encountered massive rejection by the supreme army command, so that Bruchmüller introduced it tacitly. See Zabecki 1994, 64–65 and 70.

34. Zabecki 1994, 45.

35. Bruchmüller 1922, 44.

36. Hermann Geyer elaborates: "The principle that during the attack the infantry must enter its own artillery and mortar fire, with which the assault battalions were trained with such great success, must become common to the whole infantry. This demands heedless courage and superior morale, because individual losses by friendly artillery fire must be accepted. But doing so also alleviates the close combat with the enemy infantry and their machine guns. The total losses will thus be substantially lower. This must be possible. The energy of the infantry attack and its success fundamentally depend on it" (1921, 672). See also Kittler 1996, 222.

37. See Geyer 1921, 107, and Zabecki 1994, 56.

38. For Wittgenstein, in any case, to be positioned in the eastward-directed observation post means to be granted a probability of survival of six weeks. See also the conference report by Sven Felix Kellerhoff, "Die vergessene Front. Eine Berliner Tagung rekonstruiert den anderen Krieg, der von 1914 bis 1918 den Osten Europas verwüstete" (2004).
39. Heidegger 1978, 308. See also Kittler 1996, 224.
40. Ludwig Wittgenstein, quoted in Waismann 1979, 68.
41. Wittgenstein 1991, 72.
42. Wittgenstein 1991, 73.
43. Macho 1996, 42.
44. Wittgenstein 1991, 41.
45. Wittgenstein 1991, 37.
46. Wittgenstein 1998, 25. See also Macho on Wittgenstein's visual strategies (1996, 30–34).
47. Award application in the War Archive, Vienna, quoted in Wittgenstein 1991, 141–142. The application was granted: he received the silver medal for bravery— which would not be his first nor his last honor.
48. Wittgenstein 1980, 24. See Pichler 2004, 105.
49. See Zabecki 1994, 45.
50. Ganghofer 1915, 198.
51. McGuinness 1988, 240.
52. See Kittler 1996, 210.
53. Jünger 2004, 117.
54. Jünger 2004, 117–118.
55. Jünger 2004, 125–126.
56. See Kittler 1996, 207n14.
57. See Linnenkohl 1990, 150–151. The shooting rules of the artillery state explicitly: "The opening of fire is ordered by the troop leader. Premature opening of fire betrays one's position to the enemy"(Anonymous 1923), 14.
58. Jünger 1979, 455. The scene is reminiscent of Jünger's observation of the front at the Somme, though now it is no longer grabbing a rifle on his own authority that occurs to him, but the order that the shooting rule would have required: to open fire.
59. Linnenkohl 1990.
60. See Kant 1917, 259.
61. Wittgenstein 1961, 91.
62. Wittgenstein 1961, 7.
63. See Wittgenstein 1961, 7.
64. See Wittgenstein 1961, 19.
65. Wittgenstein 1961, 7.
66. Wittgenstein 1961, 19.
67. See Wittgenstein 1961, 7, and Wright 1955, 532–533.
68. Wittgenstein 1961, 7.

69. Wittgenstein 2001, 69.
70. Wittgenstein 2001, 69.
71. Wittgenstein 1961,, 82.
72. Wittgenstein 1991, 74.
73. Wittgenstein 2001, 69.
74. Wittgenstein 1961, 2.
75. Wittgenstein 1961, 11.
76. Wittgenstein 1961, 79.
77. Wittgenstein 2001, 89.
78. Jünger, quoted in Heidegger 1958, 65.
79. Wittgenstein 1979a, 170.
80. See Wittgenstein 2001, 8.
81. Wittgenstein 1979a, 170.
82. See Syed 2001, 10. Syed supports the thesis of the great chess historian Harold J. Ruthven Murray, according to whom chess was designed as a war game at one stroke. The question arises because it displays elements that can be found in no other game. Caturanga, as the game of chess but also the military was called in India, first appears in the sixth century as the gift of a north Indian ruler in the records of the Persian royal court. Syed surmises that the unnamed ruler was the north Indian king Sarvavarman. The formation of the military into four divisions, consisting of elephants, horses, wagons, and foot soldiers—a division that also characterizes the first known forms of chess—goes back to him.
83. Wittgenstein 1979c, 104.
84. Wittgenstein 1979c, 103.
85. Weyl's corresponding article in *Symposion* is explicitly mentioned, despite Wittgenstein's usual rejection of providing the academic apparatus with citations, which prevented him from submitting a dissertation in Cambridge. Indeed, it is apparent from Waismann's notes that Wittgenstein also knew Weyl's article in the *Handbuch der Philosophie*, which presents the schematism of chess as well as Hilbert's proof procedure. See Weyl 2009, 61.
86. See Wittgenstein 1974, 50–51.
87. Waismann had been invited to the conference on short notice. However, the typescript of his contribution did not reach the publisher of the conference proceedings. See Waismann 1931, 87.
88. Mancosu 1999, 37.
89. Brouwer had been cast out of the publishing circle of the *Mathematische Annalen* the same year as Hilbert. As long as the foundational debate was fought out in such journals and from academic posts such as Hilbert's in Göttingen, there was no decisive break. Rather, this occurred only when Brouwer declined the offer of a professorship in Göttingen, let Berlin court him, and developed a center in Amsterdam that would attract other mathematicians. Ultimately, it seemed that the debate was fueled more by the expansion of institutions and less by the dissemination of intuitions.

90. See McGuinness 1979, 16.

91. Monk 1990, 249.

92. See Rudolf Carnap's statement in Hahn 1931, 141.

93. See Hahn 1931, 143.

94. See Weyl 1968b, 2:157.

95. See Heyting's statement in Hahn 1980, 146.

96. Hilbert 1967, 475.

97. Neumann 1983, 62.

98. Neumann 1983, 117.

99. Wittgenstein 1979c, 103.

100. It is noteworthy that Neumann finds support for his representation of formalism, just as Wittgenstein does, in Weyl's work.

101. vNeumann's statement in Hahn 1980, 144.

102. Gödel 1931, 150.

103. Frege already rejected a formalism that did not go beyond the status of a game: "If it [formal arithmetic] is a game with figures, then it contains theorems and proofs no more than chess does" (1903, 101). Wittgenstein, who had gone to Cambridge to work with Russell at Frege's recommendation, apparently found in this statement of Frege's the reason Hilbert was prevented from shifting onto a metalevel for the control of a rule system: "I can play with chessmen according to certain rules. But I could also invent a game in which I play with the rules themselves: Now the rules of chess are the pieces of my game and the laws of logic for instance are the rules of the game. *In this case I have yet another game and not a metagame.* What Hilbert does is mathematics and not metamathematics. It is another calculus, just like any other one" (Wittgenstein 1979b, 120–121). Hermann Weyl first made the chess analogy in his article "Randbemerkungen zu Hauptproblemen der Mathematik" (1924, 147–148). The article is a response to Hilbert's "The New Grounding of Mathematics" (1996, originally published in 1922), which Hilbert, in turn, felt challenged to undertake by Weyl's article "Über die neue Grundlagenkrise der Mathematik" (1968b, originally published in 1921). Nonetheless, Weyl belonged to the innermost circle around Hilbert in Göttingen. In subsequent replies, Hilbert took up the analogy to the formula game: See Hilbert 1967, 474–475. Meanwhile, Weyl publicized the equation of formalist mathematics with chess in forums that were not addressed solely to professional mathematicians: for example, in the journal *Symposion* (1925, 25–30) or in the *Philosophy of Mathematics and Natural Sciene* (2009, 61).

104. See Becker 1973. (The text first appeared 1927 in the same volume of Husserl's *Jahrbuch für Philosophie und phänomenologische Forschung* as Martin Heidegger's "Sein und Zeit I." Becker himself highlights that several times in his text.) On the formula and sign game, see Becker 1973, 71, 75, 76, and 166. See also Dieudonné 1962, 550–551.

105. Gauß, quoted in Becker 1973, 41.

106. Hilbert 1992, 14.

107. Du Bois-Reymond 1974, 160, 167–171.
108. Hilbert 1902, 445.
109. Hilbert 1996a, 2:1165. Though the Homeric laughter into which Hilbert then burst was not transmitted on the radio broadcast, it was nonetheless recorded by the audio technicians, according to his biographer Constance Reid.
110. See Mehrtens 1990, 108.
111. Hilbert formulates this claim most consistently in his article "Axiomatisches Denken," originally published in 1919: "Once it has become sufficiently mature for the formation of a theory, anything which can at all be the object of scientific thought succumbs to the axiomatic method and consequentially to mathematics. By penetrating into deeper levels of axioms . . . we also gain deeper insight into the essence of scientific thought and become more and more conscious of the unity of our knowledge. Under the banner of the axiomatic method, mathematics appears to be destined to a leading role in all science" (1996, 3:1115). Mehrtens speaks in this connection of Hilbert's vision as being "almost imperialist" (1990, 132).
112. Hilbert 1988, 65. See Du Bois-Reymond 1910, 195. See also Mehrtens 1990, 133.
113. See also McCarty 2004, 517–523.
114. Du Bois-Reymond 1882, 53–55.
115. Du Bois-Reymond 1882, 166.
116. See McCarty 2004, 522.
117. Du Bois-Reymond 1882, 38.
118. Du Bois-Reymond 1882, 40.
119. Du Bois-Reymond 1882, 41.
120. Weyl 1931, 147.
121. Weyl regards Hilbert and Brouwer as a third epoch in the foundational investigations. The first he discerns in Dedekind and Cantor and the second he ascribes to Russell (1924, 147).
122. Hilbert 1967, 464–465.
123. Müller 1923, 156.
124. Bernays 1923, 159.
125. See Bernays' editorial note in Hilbert 1996c, 2:1122, 1115–1134.
126. Bernays 1923, 160.
127. Hilbert 1996a, 1159–1160.
128. Hilbert 1996a, 1163.
129. Weyl states that with the looming victory of formalism "phenomenology as the fundamental philosophical science is doomed" (quoted in Mancosu 2002, 145).
130. Brouwer, quoted in Hermann Weyl 1968a, 529.
131. Brouwer 1996b, 391.
132. Brouwer's relationship to his mentors is noteworthy: despite intense confrontations, his adviser Diederik Johannes Kortweg would offer him his own

professorship in applied mathematics at the University of Amsterdam. Brouwer asked Hilbert, with whom he had a friendly connection for years, for advice beforehand as to whether he, of all people, should accept a professorship in applied mathematics or rather a vacant post for pure mathematics at a provincial university. Hilbert advised the former. See van Stigt 1990, 59.

133. van Stigt 1990, 410.

134. Brouwer, quoted in van Stigt 1990, 40.

135. To that end, he preferred to practice mathematics not from his desk, but lying with closed eyes or sitting cross-legged in his hut. This domicile maintained distance from the city of Amsterdam and his professorship there and stood open only to the smaller circle of his disciples and selected visitors—Hilbert among them. Details in van Stigt 1990, 49.

136. See van Dalen 1999, 277. Brouwer developed the memorandum into articles that appeared in several publications: (1919/1920, 311–331, 300–307). Brouwer had already previously published articles on photogrammetry.

137. See van Dalen 1999, 276–277.

138. A brief and pointed discussion of the role of survey photographs and map production at the beginning of the First World War can be found in an article by Bernhard Siegert (1996, 268–278).

139. Brouwer's photogrammetric works are not included in his complete writings. Nor have they apparently been translated elsewhere from the Dutch. Professor Jürgen Albertz, an expert in photogrammetry and well versed in its history, informed me in response to my inquiry that he had never before heard of Brouwer's contributions to photogrammetry (personal communication).

140. Hilbert 1996b, 146.

141. Weyl 1925, 1.

142. Weyl 1968b, 143.

143. Weyl 1968b, 56.

144. Hilbert 1996c, 1119 and 1132.

145. Hilbert 1983, 192.

146. Brouwer 1996a, 1181.

147. Brouwer, quoted in van Stigt 1990, 87.

148. Weyl 1968b, 45.

149. Schmitt 2005, 3.

150. Schmitt 2005, 6–7.

151. Schmitt 2005, 7.

152. Weyl 1925, 20.

153. Pachukanis 1970, 23–24.

154. Weyl 1968b, 155.

155. Weyl 1968b, 153. For the concept of the medium in Weyl's work, see Röller 2000.

156. Du Bois-Reymond 1882, 87.

157. Weyl 1968b, 166. Emphasis in the original.

158. Schmitt 1991, 11–12.

159. Schmitt 1991, 80.

160. Schmitt 1991, 76. In this connection, Schmitt elaborates: "The fact that between the self-enclosed state territory and the—if I may say so—non-state nothingness of international law there are in reality many peculiar formations that are neither purely internal nor purely external to the state, that not only the territorial sovereignty of the state but also spatial sovereignties of various kinds belong to the reality of international law, goes unrecognized in the simple either-or of inter-state in a similar way as the dualism of inter-state and a state's internal law is not at all capable of constructing overarching contexts" (66–67).

161. Bergson 2007, 255.

162. See Smith 2003, 66. In this connection, Smith is concerned with establishing the concept of the system as an all-encompassing term: "Systems in many respects resemble machines. A machine is a little system, created to perform, as well as to connect together, in reality, those different movements and effects which the artist has occasion for. A system is an imaginary machine invented to connect together in the fancy those different movements and effects which are already in reality performed" (66).

163. Galton 1880, 301–318.

164. See the illustrations in Galton 1889, 63, 107. His most famous "mechanical illustration" became known as the Galton board. It demonstrates statistical contexts such as Gaußian distributions and correlations.

165. Along with the chess schema, Weyl, for one, introduces as an analog to a formalized mathematics a machine that is later taken up as the so-called Zeus machine and that passes through all the natural numbers in a finite process of "acts of decision" by assuming that the time that is spent for a decision is constantly halved. See Weyl 1925, 22, and Weyl 2009, 42. Hans Reichenbach invokes machine guns and water hoses in order to set the mathematics of the continuum in relation to a mathematics of discrete elements (1929, 275–276).

166. See Brouwer 1981, 4. Brouwer's temporal postulation goes back to his dissertation. On the occasion of his lectures in Vienna, Brouwer further elaborated his reflections on the perception of time (*tijdsgewaarwording*). See also van Dalen 1999, 2:562–566..

167. Husserl 1970, 48.

168. Husserl 1970, 48.

169. Husserl 1970, 5.

170. Husserl 1970, 58.

171. Husserl 1970, 4.

172. It is therefore necessary to approach with caution interpretations that already seek to extract from it hints of the hardships that Husserl suffered first-hand at the hands of the National Socialist Führer state in the final years of his life.

173. Husserl 1970, 52.

174. Turing 1936–1937.

175. It is a paper machine in a double sense: a construct on paper and at the same time the hypothesis of a machine that essentially consists of an endless paper tape inscribed with a limited supply of signs according to fixed rules and in specific positions. Signs can also be erased. The procedure that follows a system of rules is known as a Turing machine; the procedure that encompasses all possible procedures is known as a universal Turing machine. See also Hodges 1983, 96–107.

176. Husserl 1970, 46. Emphasis in original.

177. Husserl 1970, 46.

178. Husserl 1970, 26.

179. Husserl 1970, 27.

180. It goes without saying that all previous machine conceptions, which saw their modus operandi in reproduction, have thereby been superseded.

181. Proceeding from Turing's work, mathematicians such as Andrey Nikolaevich Kolmogorov and Gregory J. Chaitin developed an algorithmic concept of randomness. For an introductory discussion, see Chaitin 1998, 11.

182. See Hodges 1983, 123.

183. Manzano 1997, 221.

184. Church even dispensed with the use of a typewriter and chose all his life to use paper and different-colored inks for the inscription of the abundant indices and signs of his logical calculus.

185. Husserl 1970, 4.

186. Oswald Spengler, quoted in Scholz 1920, 5. See also Spengler 1926, 1:40–41.

187. See Stock 1987, 31.

188. The only exception is a colleague at King's College, Cambridge, where Turing himself resided academically. See Hodges 1983, 123. Neither from Hermann Weyl nor from John von Neumann did Turing receive feedback on his work—he had sent it to both of them (123). At the same time, he had gone to Princeton at the recommendation of his Cambridge mentor Newman for a long scholarly residence, where—along with Alonzo Church (his advisor there)— Weyl and Neumann taught: the two mathematicians who had recently been most intensely involved in the foundational debate around the question of calculability.

189. See Hodges 1983, 152.

190. The Nazi regime could not prevent Scholz from maintaining contact to the Polish school of logicians; in particular, he supported Jan Lukasiewicz. It was no more a concession to the powers that be when Scholz awarded the Ernst Schröder Prize to the logician J. C. C. McKinsey in the United States in 1941. After the war, McKinsey belonged to the core group of America's most prominent strategic think tank, the Research and Development Corporation (RAND), though

only as long as his critique of Neumann's game theory and his homosexuality did not force him to leave the institution. McKinsey did not let that stop him from writing a standard work of game theory following Neumann. See also Mirowski 2002, 320–321.

191. Zuse 1986,76. In March 1945, Scholz reviewed quite favorably Zuse's dissertation on approaches to a theory of general calculation, which prepared the way for his algorithmic language—the plan calculus. Ultimately, however, Zuse did not submit the work as his dissertation. See Petzold 1992.

192. Zuse reports having heard of Turing's work only later. See Zuse 1972, 5.

193. See Bauer 1997, 385–393.

194. Ironically, Hasenjäger remained connected to Turing even after the war through machine configurations because as the assistant and successor to Scholz as professor for mathematical logic and foundational research at Münster, he developed numerous Turing machines of different sorts.

195. Winterbotham 1974.

196. Zuse 1972, 1.

197. Zuse 1986, 51.

198. See Zuse 1972, 35.

199. See Turing 1953, 286–295. Turing's subchapter belongs to the article *Digital Computers Applied to Games*; however, its introduction and other chapters, according to Hodges, were not written by Turing. Turing had already conceived of the first chess-playing paper machines in Bletchley Park during the Second World War. See also Shannon 1950b, 256–275; 1950a, 2124–2133.

200. See Shannon 1955, 447–453.

201. See Shannon 1950b, 637.

202. Ernst Zemolo elevated chess to the application of his axiomatic set theory at the fifth international congress of mathematics. The point of departure for Zermelo's proof is the fact that the regular game of chess knows a finite number of game constellations but grants a theoretically endless number of rule-consistent moves. Zermolo makes a claim about the determinedness with which a checkmate position, when it is possible, occurs even if the opponent can make a theoretically infinite number of moves (1913, 501–504). Two Hungarian mathematicians, László Kalmár and Dénes König, took up Zermolo's approach in 1928 and 1929. The extent to which the application of mathematics to the object of the game can be stretched is demonstrated by the fact that now solutions for chessboards with an endless number of fields and move possibilities are hypothesized. But the fact that Zermelo's approach is itself without consequences for the game of chess applies equally to their contributions. Finally, Max Euwe also joined the discussion by demonstrating Brouwer's intuitionist method by way of the same example. He reduces König's endless chessboard to a constructive dimension, because intuitionists prohibit themselves from turning existential statements into properties of uncountable elements. Max Euwe would inciden-

tally become known as a world chess champion, and as the only mathematician at the grave of L. E. J. Brouwer, he would give the eulogy. Neither König, Kalmár, nor Euwe left unmentioned that they received suggestions from a certain Dr. Neumann. See König 1927, 121–130; Kalmár 1928, 65–85; and Euwe 1929, 633–642. The sixth chapter of Euwe's book deals more extensively with the role of the game at the time of the foundational debate.

203. Neumann 1928, 298.

204. Neumann 1959, 299–300.

205. Neumann 1959, 300.

206. Although the question of priority itself does not seem to have particularly interested Borel, Maurice Fréchet took his side after his death. What is noteworthy about the confrontation fought out in 1953 in the magazine *Econometrics* is Fréchet's indication that it was Georges Guilbaud who first referred him to Borel's writings on game theory. With Guilbaud in the background, the priority debate turns out to be a symptom of the growing attention that game theory received in intellectual Paris after 1950. See Fréchet 1953, 95.

207. See Ulam 1958, 10n3.

208. The Rockefeller dynasty also financed Hilbert's new building for the mathematical institute, which would be abandoned only a few years later. See Hilbert 1971, 82.

209. Upon the publication of the work in 1928, Neumann pointed out that it had been presented "with some abridgements," which permits the assumption that his theory existed in elaborated form in 1926. The date of his talk is of special significance because his works on quantum mechanics in particular had appeared earlier, but only later emerged in Göttingen. Scholars have amply demonstrated and discussed the fact that Neumann had already become a central figure in the circle of mathematicians around David Hilbert, as he was extremely productive in several of their fields of research at the same time. It seems all the more astonishing that scarcely anything is known about Neumann's time in Göttingen, aside from his publications. In the Göttingen university archive, there are no traces of Neumann's residency, according to a statement by the archivist there. Neumann's biographer Norman Macrae surmises that he might have arrived in Göttingen in early autumn and speculates about previous residencies. See Macrae 1992, 115. Nonetheless, there is no doubt that Neumann came to Göttingen for the first time on November 12, 1926, and stayed there until July 1, 1927. The dates are supported by the Göttingen city archive, which retains the register of residents with Neumann's entry. The time period corresponds to the information in the documents of the Rockefeller Archive Center, in which Neumann's absence between the semesters is also indicated: see Neumann's letter to Trowbridge of May 12, 1927.

210. See Mirowski 2002, 105–116.

211. Ulam 1958, 11–12.

212. Neumann 1959, 306.
213. See Shannon 1949, 662–663.
214. The founding manifesto of the French group of mathematicians known by the collective pseudonym "Nicolas Bourbaki" begins with the question: "La Mathématique, ou les Mathématiques?" See Bourbaki 1948, 35.
215. See Heidegger 1982, 106.

Bibliography

Alberti, Leon Battista. 1973. Ludi Rerum Mathematicarum. Vol. 3. In Opere Volgari, ed. Cecil Grayson, 130–173. Bari: G. Laterza.

von Altrock, Konstantin. 1908. *Das Kriegsspiel. Eine Anleitung zu seiner Handhabung. Mit Beispielen und Lösungen.* Berlin: Mittler und Sohn.

Anonymous. 1828. Supplement zu den bisherigen Kriegsspiel-Regeln. Zeitschrift für Kunst, Wissenschaft und Geschichte des Krieges 13(4): 68–105.

Anonymous [Albrecht von Roon]. 1854. Zur Erinnerung an den Griesheim, gestorben als erster Commandant von Coblenz und Ehrenbreitenstein, am 1. Januar 1854. *Beiheft zum Militair-Wochenblatt:* 1–29.

Anonymous. 1869. Zum Kriegsspiel. Militair-Wochenblatt 35 and 37: 276–277, 292–295.

Anonymous. 1874. Zur Vorgeschichte des v. Reiswitz'schen Kriegsspiels. Militair-Wochenblatt 73:693–694.

Anonymous. 1923. Die Ausbildung der Artillerie auf Grund der Kampfschule und Schießvorschrift (A.B.A). Nach amtlichem Material für alle Waffen bearbeitet, zugleich 5. Beiheft zum 107. Jg. des *Militär-Wochenblatt:* 14.

Anonymous. 1936. Eine peinliche Ehrenrettung. *Das Schwarze Korps* 49 (Dec. 3): 3.

Arbib, Michael A. 1972. Toward an Automata Theory of Brains. *Communications of the ACM* 15 (7): 521–527.

Arbib, Michael A. 2000. Warren McCulloch's Search for the Logic of the Nervous System. *Perspectives in Biology and Medicine* 43 (2): 193–216.

Aschhoff, Volker. 1995. Geschichte der Nachrichtentechnik: Nachrichtentechnische Entwicklungen in der ersten Hälfte des 19. Jahrhunderts. Vol. 2. Berlin: Springer-Verlag.

de Balzac, Honoré. 1967. *Lettres à Madame Hanska.* Vol. 1. Ed. Roger Pierrot. Paris: Editions du Delta.

de Balzac, Honoré. 1981. *La Comédie Humaine*. Vol. XII. Ed. Pierre-Georges Castex. Paris: Gallimard.

Baudrillard, Jean. 1993. Symbolic Exchange and Death, trans. Iain Hamilton Grant. London: Sage Publications. Originally published as *L'échange symbolique et la mort*. Paris: Gallimard, 1976.

Baudrillard, Jean. 1994. The Precession of the Simulacra. In Simulacra and Simulation, trans. Sheila Faria Glaser, 1–42. Ann Arbor: The University of Michigan Press. Originally published as *La précéssion de simulacres*. Traverses 10 (1978): 3–37.

Bauer, Friedrich L. 1997. *Decrypted Secret. Methods and Maxims of Cryptology*. Berlin: Springer.

Becker, Oskar. 1973. *Mathematische Existenz. Untersuchungen zur Logik und Ontologie mathematischer Phänomene*. Tübingen: Niemeyer.

Beer, Stafford. 1993. World in Torment: A Time Whose Idea Must Come. *Kybernetes* 22 (1): 15–43.

Benjamin, Walter. 1997. Letter to Schmitt from Dec. 9, 1930. Vol. 3. Gesammelte Briefe, ed. Christoph Gödde and Henri Lonitz, 558. Frankfurt: M. Suhrkamp. English translation by Samuel Weber in "Taking Exception to Decision: Walter Benjamin and Carl Schmitt," *Diacritics* 22: 3/4: 5 (1992).

Bergmann, Werner. 1985. *Innovationen im Quadrivium des 10. und 11. Jahrhunderts. Studien zur Einführung von Astrolab und Abakus im lateinischen Mittelalter. Sudhoffs Archiv*. Vol. 26. Stuttgart: Steiner-Verl.-Wiesbaden.

Bergson, Henri. 2007. Matter and Memory, trans. Nancy Margaret Paul and W. Scott Palmer. New York: Cosimo Classics. Originally published as *Matiere et memoire*. Paris: Presses Universitaires de France, 1896.

Bernays, Paul. 1923. Erwiderung auf die Note von Herrn Aloys Müller: "Zahlen als Zeichen." *Mathematische Annalen* 90:159–163.

Bernays, Paul. 1976. Probleme der theoretischen Logik. In Abhandlungen zur Philosophie der Mathematik, 1–16. Darmstadt: Wissenschaftliche Buchgesellschaft.

Biermann, Kurt-Reinhard. 1958/1959. Zum Verhältnis zwischen Alexander von Humboldt und Carl Friedrich Gauß. Wissenschaftliche Zeitschrift der Humboldt-Universität zu Berlin, Mathem.-Naturw. *Reihe* 8 (1): 121–130.

Biermann, Kurt-Reinhard, ed. 1959. *Johann Peter Gustav Lejeune Dirichlet. Dokumente für sein Leben und Wirken. Abhandlungen der Deutschen Akademie der Wissenschaften zu Berlin. Klasse für Mathematik, Physik und Technik 2*. Berlin: Akademie-Verlag.

Biermann, Kurt-Reinhard, ed. and trans. from French. 1982. Briefwechsel zwischen Alexander von Humboldt und Peter Gustav Lejeune Dirichlet. Berlin: Akademie-Verlag.

Biermann, Kurt-Reinhard. 1988. Die Mathematik und ihre Dozenten an der Berliner Universität. 1810–1933. Stationen auf dem Wege eines mathematischen Zentrums von Weltrang. Berlin: Akademie-Verlag.

Bischoff, Bernhard. 1967. Das griechische Element in der abendländischen Bildung des Mittelalters. In Mittelalterliche Studien. Ausgewählte Aufsätze zur Schriftkunde und Literaturgeschichte. Vol. 2, 246–275. Stuttgart: Hiersemann-Verlag.

Blumenberg, Hans. 1990. Doppelte Buchführung. Synopse der Kriegstagebücher Wittgensteins 1914–1916. Frankfurter Allgemeine Zeitung 96 (May 25):N3.

Booß-Bavnbek, Bernhelm, and Jens Høyrup, eds. 2003. Mathematics and War. Basel: Birkhäuser Verlag.

Borst, Arno. 1986. Das mittelalterliche Zahlenkampfspiel. Supplemente zu den Sitzungsberichten der Heidelberger Akademie der Wissenschaften: Philosophisch-historische, Klasse, Vol. 5. Heidelberg: Carl Winter Universitätsverlag.

Borst, Arno. 1990. Rithmimachie und Musiktheorie. In Geschichte der Musiktheorie: Rezeption des antiken Fachs im Mittelalter. Vol. 3. Ed. Frieder Zaminer, 253–288. Darmstadt: Wissenschaftliche Buchgesellschaft.

Bosse, Heinrich. 1990. Der geschärfte Befehl zum Selbstdenken. Ein Erlaß des Ministers v. Fürst an die preußischen Universitäten im Mai 1770. In Institution Universität, Diskursanalysen 2, ed. Friedrich Kittler, Manfred Schneider, and Samuel Weber, 31–62. Opladen: Westdeutscher Verlag.

Bourbaki, Nicolas. 1948. L'architecture des mathématiques. In Les grands courants de la pensée mathématique, L'Humanisme scientifique de demain. Vol. 1. Ed. Francois le Lionnais, 35–47. Paris: Cahiers du sud.

Bramer, Benjamin. 1630. Beschreibung eines sehr leichten Perspectiv und grundreissenden Instruments auff einem Stande. Auff Johan Faulhabers weitere Continuation seine mathematischen Kunstspiegels geordnet. Frankfurt.

Bramer, Benjamin. 1648. Bericht zu M. Jobsten Burgi seligen Geometrischen Triangular Instruments. Mit schönen Kupfferstücken hierzu geschnitten. Kassel.

Brandes, Georg. 1902. Feldmarschall Moltke. In Brandes. Vol. 1, 11–36. München: Gesammelte Schriften. Deutsche Persönlichkeiten.

Bredekamp, Horst. 2004. Die Fenster der Monade. Gottfried Wilhelm Leibniz' Theater der Natur und Kunst. Acta humaniora. Schriften zur Kunstwissenschaft und Philosophie. Berlin: Akademie Verlag.

Brouwer, Luitzen Egbertus Jan. 1919/1920. Luchtvaart en Photogrammetrie. Nieuw Tijdschrift voor Wiskunde 7:311–331, 8:300–307.

Brouwer, Luitzen Egbertus Jan. 1981. Historical Introduction and fundamental notions. In *Brouwer's Cambridge lectures on intuitionism*, ed. Dirk van Dalen, 1–20. Cambridge: Cambridge University Press.

Brouwer, Luitzen Egbertus Jan. 1996a. Mathematics, Science, and Language. In *From Kant to Hilbert. A Source Book in the Foundations of Mathematics*, ed. and trans. William B. Ewald, 1170–1185. Oxford: Clarendon Press. Originally published as Mathematik, Wissenschaft und Sprache. *Monatshefte für Mathematik und Physik* 36: 153–164, 1928.

Brouwer, Luitzen Egbertus Jan. 1996b. Life, Art, and Mysticism. Trans. Walter P. van Stigt. Notre Dame Journal of Formal Logic 36 (3): 389–429.

Bruchmüller, Georg. 1922. *Die deutsche Artillerie in den Durchbruchschlachten des Weltkrieges*. Berlin: E.S. Mittler & Sohn.

Brühl, Reinhard. 1973. *Militärgeschichte und Kriegspolitik: zur Militärgeschichtssch-reibung des preußisch-deutschen Generalstabes 1816–1945, Schriften des Militärgeschich-tlichen Instituts der Deutschen Demokratischen Republik*. Berlin: Militärverlag der DDR.

Bruhns, Karl, ed. 1877. *Briefe zwischen A. v. Humboldt und Gauß. Zum hundertjährigen Geburtstage von Gauß am 30. April 1877*. Leipzig: W. Engelmann.

von Bülow, Heinrich Dietrich. 1799. *Geist des neuern Kriegssystems: hergeleitet aus dem Grundsatze einer Basis der Operationen / auch für Laien in der Kriegskunst faßlich vorge-tragen von einem ehemaligen Preußischen Offizier*. Hamburg: Hofmann.

Busch, Oliver. 1998. *Logos syntheseos. Die euklidische Sectio canonis, Aristoxenos und die Rolle der Mathematik in der antiken Musiktheorie. Veröffentlichungen / Staatliches Institut für Musikforschung Preußischer Kulturbesitz*. Vol. 10. Berlin: Staatl. Inst. für Musikforschung Preußischer Kulturbesitz.

Busche, Hubertus. 1997. *Leibniz' Weg ins perspektivische Universum. Eine Harmonie im Zeitalter der Berechnung. Paradeigmata*. Vol. 17. Hamburg: Meiner.

Butzer, Manfred Jansen, and Hubert Zilles. 1984. Zum bevorstehenden 125. Todestag des Mathematikers Johann Peter Gustav Lejeune Dirichlet (1805–1859). Mitbe-gründer der mathematischen Physik im deutschsprachigen Raum. Sudhoffs Archiv 68 (1): 1–20.

Cantor, Moritz. 1922. *Vorlesungen über Geschichte der Mathematik*. Leipzig: Teubner.

Carsten, Francis L. 1966. The Reichswehr and Politics: 1918–1933. Oxford: Claren-don Press. Originally published as *Reichswehr und Politik. 1918–1933*. Köln, Berlin: Kiepenheuer & Witsch Verlag, 1964.

Cassirer, Ernst. 1921. Heinrich von Kleist und die Kantische Philosophie. In Idee und Gestalt, 157–202. Berlin: B. Cassirer.

Chaitin, Gregory J. 1998. The Limits of Mathematics. A Course on Information Theory and the Limits of Formal Reasoning. Singapore: Springer.

von Clausewitz, Carl. 1966. Meine Vorlesungen über den kleinen Krieg, gehalten auf der Kriegs-Schule 1810 und 1811. Schriften, Aufsätze, Studien, Briefe. Dokumente aus dem Clausewitz-, Scharnhorst- und Gneisenau-Nachlaß sowie aus öffentlichen und privaten Sammlungen. Ed. Werner Hahlweg, 226–449. Deutsche Geschichtsquellen des 19 und 20. Jahrhunderts, Vol. 49. Göttingen: Vandenhoeck und Ruprecht.

von Clausewitz, Carl. 1980. Die wichtigsten Grundsätze des Kriegführens zur Ergänzung meines Unterrichts bei Sr. Königlichen Hoheit dem Kronprinzen. In Vom Kriege. Hinterlassenes Werk des Generals Carl von Clausewitz, ed. Werner Hahlweg, 1047–1086. Bonn: Dümmlers.

von Clausewitz, Carl. 2007. On War, trans. Michael Howard and Peter Paret. Oxford: Oxford University Press. Originally published as *Vom Kriege*. Bonn: Dümmlers Verlag, 1832.

von Cochenhausen, Friedrich. 1934a. Anleitung für die Anlage und Leitung von Planübungen und Kriegsspielen. Berlin: Schröder.

von Cochenhausen, Friedrich. 1934b. Conrad von Hoetzendorf. Eine Studie über seine Persönlichkeit. Schriften der kriegsgeschichtlichen Abteilung im historischen Seminar der Friedrich-Wilhelms-Universität Berlin. Berlin: Junker und Dünnhaupt.

von Cochenhausen, Friedrich. 1936a. Einleitung [Introduction]. In Die wichtigsten Grundsätze des Kriegsführens by Carl von Clausewitz 5–8. Berlin: Junker und Dünnhaupt.

von Cochenhausen, Friedrich. 1936b. Führertum. Vol. 1. Handbuch der neuzeitlichen Wehrwissenschaften. Herausgegeben im Auftrage der Deutschen Gesellschaft für Wehrpolitik und Wehrwissenschaften und unter Mitarbeit umstehend aufgeführter Sachverständiger von Hermann Franke: Wehrpolitik und Kriegsführung, 102–103. Berlin: Walter de Gruyter.

Cohen, Kalman J., and Eric Rhenman. 1961. The Role of Management Games in Education and Research. Management Science 7 (2): 131–166.

Dannhauer, Ernst Heinrich. 1874. Das Reiswitzsche Kriegsspiel von seinem Beginn bis zum Tode des Erfinders 1827. Militair-Wochenblatt 56:527–532.

Dieudonné, Jean. 1962. Les méthodes axiomatiques modernes et les fondements des mathématiques. In Les Grands Courants de la Pensée mathématique, ed. F. Le Lionnais, 543–555. Paris: Blanchard.

Dirks, Carl, and Karl-Heinz Janßen. 1999. Der Krieg der Generäle. Hitler als Werkzeug der Wehrmacht. Berlin: Propyläen.

Domarus, Max. 1965. Hitler. Reden und Proklamationen 1932–1945. Vol. 2. Munich: Süddeutscher Verlag, 1965.

von Domaszewski, Alfred. 1885. Die Fahnen im römischen Heere. Abhandlungen des archäologisch-epigraphischen Seminars der Universität Wien. Vol. 5, no. 5. Wien: Verlagsbuchhandlung Carl Gerold's Sohn.

Du Bois-Reymond, Émile. 1974. Die sieben Welträtsel. In der Leibniz-Sitzung der Akademie der Wissenschaften am 8. Juli 1880 gehaltene Rede. In Vorträge über Philosophie und Gesellschaft, ed. Siegried Wollgast, 159–187. Hamburg: Meiner.

Du Bois-Reymond, Paul. 1882. Die Allgemeine Funktionentheorie: Metaphysik und Theorie der mathematischen Grundbegriffe: Grösse, Grenze, Argument und Function, erster Teil. Tübingen: Laupp.

Du Bois-Reymond, Paul. 1910. Was will die Mathematik und der Mathematiker? Rede beim Antritt der ordentlichen Professur der Mathematik an der Universität Tübingen (1874) gehalten. Jahresbericht der deutschen Mathematiker-Vereinigung 19:190–198.

Edgerton, Samuel Y. 1980. The Renaissance Artist as Quantifier. In The Perception of Pictures. Vol. 1. Ed. Margaret A. Hagen, 179–212. New York: Academic Press.

Erdmann, Carl. 1935. Die Entstehung des Kreuzzugsgedankens. Stuttgart: Verlag W. Kohlhammer.

Euwe, Max. 1929. Mengentheoretische Betrachtungen über das Schachspiel. *Proceedings. Koninklijke Akademie van Wetenschappen Te Amsterdam* 32. Presented by Prof. R. Weitzenböck. No. 5: 633–642.

Faber, Marion. 1988. Das Schachspiel in der europäischen Graphik (1550–1700), Wolfenbütteler Arbeiten zur Barockforschung 15. Wiesbaden: Harrassowitz.

Fangohr, Friedrich-Joachim. 1951. "Beitrag zur Studie ueber Zweck und Art der Durchfuehrung von Kriegsspielen, Planuebungen usw. im deutschen Heer," attachment 1 of "Über 'Kriegsspiele.'" Bundesarchiv-Militärarchiv, Bestand: P-094: 131, 133, 135.

Faulhaber, Johann. 1610. *New erfunden Instrument zu den Irregular Fortification*. Ulm.

Faulhaber, Johann. 1644. Mechanische Reißladen. Augsburg.

Finker, Kurt. 1994. Der 20. Juli 1944. Militärputsch oder Revolution? Berlin: Dietz-Verlag.

Fleckenstein, Josef. 1980. Die Rechtfertigung der geistlichen Ritterorden nach der Schrift "De laude novae militiae" Bernhards von Clairvaux. In Die geistlichen Ritterorden Europas, ed. Josef Fleckenstein and Manfred Hellmann, 9–22. Sigmaringen: Thorbecke Jan Verlag.

Foucault, Michel. 2003. Society Must be Defended: Lectures at the Collège de France, 1975–1976, ed. Mauro Bertani and Alessandro Fontana, trans. David Macey. New York: Picador. Originally published as Il faut défendre la société: cours au Collège de France (1975–1976). Paris: Éditions de Seuil/Gallimard, 1997.

Fréchet, Maurice. 1953. Emile Borel, Initiator of the Theory of Psychological games and its Application. Econometrica 21:95–96.

Frege, Gottlob. 1903. Grundgesetze der Arithmetik II. Jena: Pohle.

Freud, Sigmund. 1940. Massenpsychologie und Ich-Analyse. In Gesammelte Werke. Vol. 13. Ed. Anna Freud and Edward Bibring. Frankfurt am Main: Fischer.

Friedlein, Gottfried. 1863. Das Rechnen mit Columnen vor dem 10. Jahrhundert. Zeitschrift für Mathematik und Physik 9:297–330.

Furttenbach d. Ältere, Joseph. 1663. Mannhafter Kunst-Spiegel. Augsburg: Verlag Hans Schultes der Jüngere.

Galison, Peter. 1994. The Ontology of the Enemy: Norbert Wiener and the Cybernetic Vision. Critical Inquiry 21:228–266.

Galton, Francis. 1889. Natural Inheritance. London: Macmillan and Co.

Galton, Francis. 1880. Statistics of Mental Imagery. Mind 5 (19): 301–318.

Ganghofer, Ludwig. 1915. Reise zur deutschen Front 1915. Berlin: Ullstein & Co.

Genuth, Joel. 1988. Microwave Radar, the Atomic Bomb, and the Background to U.S. Research Priorities in World War II. Science, Technology & Human Values 13:276–289.

Geyer, Hermann. 1921. Der Angriff im Stellungskrieg. In Urkunden der Obersten Heeresleitung über ihre Tätigkeit 1916/1918, ed. Erich Ludendorff, 672. Berlin: E. S. Mittler und Sohn.

Gödel, Kurt. 1931. Nachtrag [Postscript]. Erkenntnis. Bericht über die 2, Tagung für Erkenntnislehre der exakten Wissenschaften in Königsberg (2/3): 147–151.

Gow, James. 1884. Short History of Greek Mathematics. Cambridge, Mass.: University Press.

Groener, Wilhelm, ed. 1930. Führertum. 25 Lebensbilder von Feldherren aller Zeiten, Auf Veranl. d. Reichswehrmin, Dr. Groener, bearb. v. Offizieren d. Reichsheeres und zusammengestellt von Generalleutnant von Cochenhausen. Berlin: E. S. Mittler & Sohn.

Grüger, Gert, and Jörg Schnadt. 2000. Die Entwicklung der geodätischen Grundlagen für die Kartographie und die Kartenwerke 1810–1945. In Berlin-Brandenburg im Kartenbild, ed. Wolfgang Scharfe and Holger Scheerschmidt, 113–136. Berlin: Staatsbibliothek zu Berlin.

Hahlweg, Werner. 1973. Die Heeresreform der Oranier. Das Kriegsbuch des Grafen Johann von Nassau-Siegen, ed. Historische Kommission für Nassau. Wiesbaden: Histor. Komm.

Hahn, Hans. 1980. Discussion about the Foundation of Mathematics. (Contributions by Hahn, Carnap, von Neumann, Scholz, Heyting, Gödel, Reidemeister.) In Empiricism, Logic, and Mathematics: Philosophical Papers, ed. Brian McGuinness, trans. Hans Kaal, 31–38. Dordrecht: D. Reidel Pub. Material not included in that collection is from original German source: Diskussion zur Grundlegung der Mathematik, Erkenntnis: Bericht über die 2, Tagung für Erkenntnislehre der exakten Wissenschaften in Königsberg 2 (2–3) (1931): 135–145.

Harnack, Adolf. 1900. Geschichte der Königlich Preussischen Akademie der Wissenschaft. Vol. 1, first part. Berlin: Reichsdruckerei.

Harsdörffer, Georg Philipp. 1990. Delitiæ Mathematicæ et Physicæ. Der Mathematischen und Philosophischen Erquickstunden Zweyter Teil. Neudruck der Ausgabe Nürnberg 1651. In Texte der Frühen Neuzeit, ed. Jörg Jochen Berns. Frankfurt am Main: Keip.

Hartlaub, Felix. 2002. In den eigenen Umriss gebannt. Kriegsaufzeichnungen, literarische Fragmente und Briefe aus den Jahren 1939 bis 1945, ed. Gabriele Lieselotte Ewenz. Frankfurt am Main: Suhrkamp.

Hassel, Paul. 1905. Joseph Maria von Radowitz. 1797–1848. Vol. 1., 24. Berlin: Mittler.

Hausrath, Alfred H. 1971. Venture Simulation in War, Business, and Politics. New York: McGraw-Hill.

Heidegger, Hermann. 1961. Kann Kriegsgeschichtsunterricht heute noch einen praktischen Nutzen haben? Wehrkunde 10:195–199.

Heidegger, Martin. 1958. The Question of Being, trans. William Kluback and Jean T. Wilde. New York: Twayne Publishers. Originally published as Über die "Linie." In Freundschaftliche Begegnungen, ed. Armin Mohler. Frankfurt am Main: Vittorio Klostermann, 1955.

Heidegger, Martin. 1982. The Nature of Language. In On the way to language, trans. Peter D. Hertz, 57–108. San Francisco: Harper & Row. For original text, see German Das Wesen der Sprache. Dritter Vortrag am 7. Februar 1958 im Studium generale. In Unterwegs zur Sprache, 196–216. Stuttgart: Neske, 1993.

Heidegger, Martin. 1993. Sein und Zeit. Tübingen: Max Niemeyer Verlag. For an English transtation, see *Being and Time*, trans. John MacQuarrie and Edward Robinson. Oxford: Blackwell, 1978.

Heidegger, Martin. 1997. Der Satz vom Grund. In Gesamtausgabe. Pt. 1, vol. 10. Ed. Petra Jaeger. Frankfurt am Main: Vittorio Klostermann. For an English translation, see *The Principle of Reason*, trans. Reginald Lilly. Bloomington: Indiana University Press, 1991.

Henniger-Voss, Marie J. 2002. How the "New Science" of Cannons Shook up the Aristotelian Cosmos. Journal of the History of Ideas 63:371–397.

Hensel, Kurt. 1988. Gedächtnisrede auf Ernst Eduard Kummer. In Nachrufe auf Berliner Mathematiker des 19. Jahrhunderts. C.G.J. Jacobi, P.G.L. Dirichlet, E.E. Kummer, L. Kronecker, K. Weierstrass, ed. Hans Reichardt, 75–111. Leipzig: B. G. Teubner.

Hilbert, David. 1992. Die übliche Auffassung von der Mathematik und ihre Widerlegung. In Natur und mathematisches Erkennen. Vorlesungen gehalten 1919–1920 in Göttingen, ed. David Rowe, 3–35. Basel: Birkhäuser.

Hilbert, David. 1902. Mathematical Problems, trans. Mary F. Winston. Bulletin of the American Mathematical Society 8 (10): 437–479. Originally published as Mathematische Probleme, *Archiv für Mathematik und Physik* R. 3, 1, 1901: 44–63.

Hilbert, David. 1967. The Foundations of Mathematics. In From Frege to Gödel. A Source Book in Mathematical Logic, 1879–1931, ed. Jean van Heijenoort, 464–479. Cambridge, Mass.: Harvard University Press. Originally published as Die Grundlagen der Mathematik. *Abhandlungen aus dem Seminar der Hamburgischen Universität* 6 (1928): 65–85.

Hilbert, David. 1971. Über meine Tätigkeit in Göttingen. In Hilbert. Gedenkband, ed. Kurt Reidemeister, 78–82. Berlin: Springer.

Hilbert, David. 1983. On the Infinite. In Philosophy of Mathematics: Selected Readings, ed. Paul Benacerraf and Hilary Putnam, trans. Erna Putnam and Gerald J. Massey, 183–201. Cambridge: Cambridge University Press. Originally published as Über das Unendliche. *Mathematische Annalen* 95 (1926): 161–190.

Hilbert, David. 1988. Wissen und mathematisches Denken. Vorlesungen von Prof. D. Hilbert, W.S. 1922/23 (Typescript edited by W. Ackermann) Bibiliothek des Mathematischen Seminars der Univ. Göttingen.

Hilbert, David. 1996a. Logic and the Knowledge of Nature. In From Kant to Hilbert: A Source Book in the Foundations of Mathematics, Vol. 2, ed. and trans. William B. Ewald, 1157–1165. Oxford: Clarendon Press. Originally published as Naturerkennen und Logik. *Die Naturwissenschaften* 18 (1930): 959–63.

Hilbert, David. 1996b. Axiomatic Thought. In From Kant to Hilbert: A Source Book in the Foundations of Mathematics, Vol. 3, ed. William B. Ewald, 1105–1115. Oxford: Clarendon Press. Originally published as Axiomatisches Denken. *Mathematische Annalen* 78 (1918): 405–415.

Hilbert, David. 1996c. The New Grounding of Mathematics. First Report. In From Kant to Hilbert. A Source Book in the Foundations of Mathematics, ed. and trans. William B. Ewald, 1115–1134. Oxford: Clarendon Press. Originally published as Neubegründung der Mathematik. Erste Mitteilung. *Abhandlungen aus dem Mathematischen Seminar der Hamburgischen Universität*, 1 (1922): 157–177.

Hildebrand, Karl Friedrich. 1990. Die Generale der deutschen Luftwaffe: 1935–1945; die militärischen Werdegänge der Flieger-, Flakartillerie-, Fallschirmjäger-, Luftnachrichten- und Ingenieur-Offiziere einschliesslich der Ärzte, Richter, Intendanten und Ministerialbeamten im Generalsrang. Vol. 1. Osnabrück: Biblio.

Hodges, Andrew. 1983. Alan Turing: The Enigma. New York: Simon and Schuster.

Hoffmann, Peter. 2007. Oberst i. G. Henning von Tresckow und die Staatsstreichpläne im Jahr 1943. Vierteljahrshefte für Zeitgeschichte 55(2):343–344.

Hofmann, Rudolf. 1951. Über "Kriegsspiele," 1951. Bundesarchiv-Militärachiv, Bestand: P-094. Available in English: Trans. P. Nuetzkendorf, *War Games*, 1952, U.S. Army Historical Document MS P-094, Department of the Army, Office of the Chief of Military History. The manuscript is also published as: Rudolf Hofmann et al., War Games: Administrative and Technical Problems in the Conduct of All Types of Staff and Command Post Exercises. In World War II German Military Studies: A Collection of 213 Special Reports on the Second World War Prepared by Former Officers of the Wehrmacht for the United States Army: A Garland Series, Vol. 21, ed. Donald S. Detwiler. New York: Garland.

Hohrath, Daniel. 2000. Prolegomena zu einer Geschichte des Kriegsspiels. In Das Wichtigste ist der Mensch. Festschrift für Klaus Gerteis zum 60. Geburtstag, ed. Angela Giebmeyer and Helga Schnabel-Schüle, 139–152. Mainz: Kliomedia.

Horst, Michael. 1999. Wie bewahrt man eine arbeitsfähige Präsidialregierung vor der Obstruktion eines arbeitsunwilligen Reichstages mit dem Ziel "die Verfassung zu wahren" bzw. zu retten?, quoted in Wolfram Pyta, Konstitutionelle Demokratie statt monarchischer Restauration, 438.

Hossbach, Friedrich. 1954. Einflüsse Immanuel Kants auf das Denken preußischdeutscher Offiziere. Bohnenrede, gehalten am 22. April 1953 vor der Gesellschaft der Freunde Kants in Göttingen. In Jahrbuch der Albertus-Universität zu Königsberg/Pr.. Vol. IV, 139–145. Kitzingen/M.: Holzner.

Huber, Ernst Rudolf. 1937. Verfassungsrecht des Großdeutschen Reiches. Hamburg: Hanseatische Verlagsanstalt.

Huber, Ernst Rudolf. 1988. Carl Schmitt in der Reichskrise der Weimarer Endzeit. In *Complexio Oppositorum. Über Carl Schmitt*, ed. Helmut Quaritsch, 33–50. Berlin: Duncker & Humblot.

Huffman, Carl A. 1993. *Philolaus of Croton, Pythagorean and Presocratic*. Cambridge: Cambridge University Press.

Husserl, Edmund. 1970. *The Crisis of European Sciences and Transcendental Phenomenology: An Introduction to Phenomenological Philosophy*, trans. David Carr. Evanston, IL: Northwestern University Press. For the original text, see *Die Krisis der europäischen Wissenschaften und die transzendentale Phänomenologie. Eine Einleitung in die phänomenologische Philosophie*. Vol. 6, Gesammelte Werke, ed. Walter Biemel, 1–276. Haag: M. Nijhoff, 1954.

Illmer, Detlef, Nora Gädecke, Elisabeth Henge, Helene Pfeiffer, and Monika Spicker-Beck. 1987. *Rhythmomachia. Ein uraltes Zahlenspiel neu entdeckt*. München: Hugendubel.

Jacobi, Carl Gustav Jacob. 1891. Über die Pariser Polytechnische Schule. Vol. 7. Gesammelte Werke, ed. Karl Weierstrass, 355–370. Berlin: Reimer.

Jähns, Max. 1890–1891. *Geschichte der Kriegswissenschaften vornehmlich in Deutschland. XVII. und XVIII. Jahrhundert bis zum Auftreten Friedrichs des Großen 1740*. Vol. 2. Geschichte der Wissenschaften in Deutschland Neuere Zeit 22. München: Oldenbourg.

Jünger, Ernst. 1979. Kaukasische Aufzeichnungen. In *Sämtliche Werke: Strahlungen II, Tagebücher II*. Pt. 1, vol. 2, 407–492. Stuttgart: Klett-Cotta.

Jünger, Ernst. 2004. *Storm of steel*, trans. Michael Hofmann. New York: Penguin Books. Originally published as *Stahlgewittern, Tagebuch eines Stoßtruppführers*. Hannover, 1920.

Kalmár, László. 1928. Zur Theorie der abstrakten Spiele. *Acta Litterarum ac Scientiarum. Sectio Scientiarum Mathematicarum* 8 (15): 65–85.

Kant, Immanuel. 1996. What Does It Mean to Orient Oneself in Thinking? In *Religion and Rational Theology*, ed. and trans. Allen W. Wood and George di Giovanni. The Cambridge Edition of the Works of Immanuel Kant, 1–18. Cambridge: Cambridge University Press. Originally published as Was heißt: Sich im Denken orientieren? *Berlinische Monatsschrift*, October (1786): 304–330.

Kant, Immanuel. 2006. *Anthropology from a Pragmatic Point of View*, ed. and trans. Robert B. Louden. Cambridge: Cambridge University Press. For the original text, see Anthropologie in pragmatischer Hinsicht. In *Kant's gesammelte Schriften*. Vol. 7, pt. 1, ed. Königlich Preußischen Akademie der Wissenschaften, 117–333. Berlin: Vereinigung Wissenschaftlicher Verleger, 1917.

Kellerhoff, Sven Felix. 2004. Die vergessene Front. Eine Berliner Tagung rekonstruiert den anderen Krieg, der von 1914 bis 1918 den Osten Europas verwüstete, in *Die Welt*, June 2.

Kennedy, Ellen. 1988. Carl Schmitt und Hugo Ball. Ein Beitrag zum Thema "Politischer Expressionismus." Zeitschrift fur Politik 35:143–162.

Keynes, Maynard. 1921. Treatise on Probability. London: Macmillan.

Kirchmann, Kay. 1998. Blicke aus dem Bunker. Paul Virilios Zeit- und Medientheorie aus der Sicht einer Philosophie des Unbewußten. Stuttgart: Verl. Internat. Psychoanalyse.

Kittler, Friedrich. 1991. Dichter—Mutter—Kind. Munich: Fink.

Kittler, Friedrich. 1996. Il fiore delle truppe scelte. In Der Dichter als Kommandant. D'Annunzio erobert Fiume, ed. Hans Ulrich Gumbrecht, Friedrich Kittler, and Bernhard Siegert, 205–225. Munich: Wilhelm Fink.

Kittler, Wolf. 1988. Militärisches Kommando und tragisches Geschick. Zur Funktion der Schrift im Werk des preußischen Dichters Heinrich von Kleist. In Heinrich von Kleist. Studien zu Werk und Wirkung, ed. Dirk Grathoff, 56–68. Opladen: VS Verlag für Sozialwissenschaften.

Klein, Felix. 1904. Über die Aufgaben und die Zukunft der philosophischen Fakultät. Jahresbericht der deutschen Mathematiker-Vereinigung 13 (5): 267–276.

Klein, Jacob. 1936. Die griechische Logistik und die Entstehung der Algebra. Quellen und Studien zur Geschichte der Mathematik, Astronomie und Physik 3:18–105, 122–235.

von Kleist, Heinrich. 1961. Sämtliche Werke und Briefe. Vol. II. Ed. Helmut Sembdner. Darmstadt: Hanser.

Kling, Thomas. 2005. Auswertung der Flugdaten. Köln: DuMont.

Knobloch, Eberhard. 1973–1976. Die mathematischen Studien von G. W. Leibniz zur Kombinatorik. Studia Leibnitiana, Supplementa 11. Wiesbaden: Franz Steiner Verlag.

Knobloch, Eberhard. 1989. Musik. In Maß, Zahl und Gewicht. Mathematik als Schlüssel zu Weltverständnis und Weltbeherrschung, ed. Menso Folkerts, Eberhard Knobloch, and Karin Reich. Ausstellungskataloge der Herzog August Bibliothek, no. 60. Weinheim: VCH, Acta humaniora.

König, Dénes. 1927. Über eine Schlussweise aus dem Endlichen ins Unendliche. *Acta Litterarum ac Scientiarum*. Sectio Scientiarum Mathematicarum 2 (26): 121–130.

Krämer, Sybille. 1988. Symbolische Maschinen. Die Idee der Formalisierung in geschichtlichem Abriß. Darmstadt: Wissenschaftliche Buchgesellschaft.

von Krosigk, Lutz Graf Schwerin. 1989. Tagebuchaufzeichnung des Reichsfinanzministers über den Verlauf der Ministerbesprechung vom 2. Dezember 1932, 9 Uhr. Vol. 2. Akten der Reichskanzlei: Das Kabinett von Papen 1. Juni bis 3. Dezember 1932, ed. Karl-Heinz Minuth, 1037. Boppard am Rhein: H. Boldt.

Kummer, Ernst Eduard. 1988. Gedächtnisrede auf Gustav Peter Lejeune-Dirichlet. In Nachrufe auf Berliner Mathematiker des 19. Jahrhunderts: C.G.J. Jacobi, P.G.L. Dirichlet, E.E. Kummer, L. Kronecker, K. Weierstrass, ed. Hans Reichardt. Teubner-Archiv zur Mathematik, Vol. 10, 36–71. Leipzig: B. G. Teubner.

Lampe, Emil. 1906. Dirichlet als Lehrer der Allgemeinen Kriegsschule. Naturwissenschaftliche Rundschau XXI (38): 482–485.

Lange, Sven. 2003. "Der große Schritt vom Wissen zum Können"—die "applikatorische Methode" in der amtlichen Kriegsgeschichtsschreibung des Kaiserreichs. In Terra et Mars: Aspekte der Landes- und Militärgeschichte; Festschrift für Eckardt Opitz zum 65. Geburtstag, ed. Michael Busch, 218–239. Neumünster: Wachholtz Verlag.

Leibniz, Gottfried Wilhelm. 1840. Zufällige Gedanken von der Erfindung nützlicher Spiele. Aus den mündlichen Unterredungen aufgezeichnet von J. F. Feller. Vol. 2. Leibniz's Deutsche Schriften, ed. Gottschalk Eduard Guhrauer. Berlin: Veit.

Leibniz, Gottfried Wilhelm. 1880. Dissertatio de Arte Combinatoria. In Die philosophischen Schriften. Vol. 4, pt. 2. Ed. Carl Immanuel Gerhardt, 15–104. Berlin: Weidmann.

Leibniz, Gottfried Wilhelm. 1887. Letter to Pierre Rémond de Montmort. In Die philosophischen Schriften, ed. Carl Immanuel Gerhardt. Berlin: Weidmann.

Leibniz, Gottfried Wilhelm. 1986a. Agenda. Vol. 3, ser. 4. Sämtliche Schriften und Briefe, ed. Akademie der Wissenschaften der DDR, 894–902. Berlin: Akademie Verlag.

Leibniz, Gottfried Wilhelm. 1986b. Gedanken zum Entwurf der teutschen Kriegsverfassung. In Sämtliche Schriften und Briefe. Ser. 4, vol. 3. Ed. Akademie der Wissenschaften der DDR, 577–593. Berlin: Akademie Verlag.

Leibniz, Gottfried Wilhelm. 2009. Annotatio de quibusdam Ludis; inprimis de Ludo quoddam Sinico, differentiaque Scachici & Latrunculorum, & novo genere Ludi Navalis. *Miscellanea Berolinensia* (1710): 22–26. Translation by Richard J. Pulskamp, Department of Mathematics & Computer Science, Xavier University, Cincinnati, OH. December 9. Available at <http://www.cs.xu.edu/math/Sources/Leibniz/sinica-latin-english.pdf>.

Linnenkohl, Hans. 1990. Vom Einzelschuß zur Feuerwalze. Der Wettlauf zwischen Technik und Taktik im Ersten Weltkrieg. Koblenz: Bernard & Graefe.

List, Wilhelm. 1951. Beitrag zu einer Abhandlung ueber den Zweck und die Art der Durchfuehrung von Kriegsspielen im deutschen Heer. Attachment 2 of Rudolf Hofman, Über "Kriegsspiele." Bundesarchiv-Militärachiv, Bestand P-094:148.

Löbel, Uwe. 1992. Neue Forschungsmöglichkeiten zur preußisch-deutschen Heeresgeschichte. Zur Rückgabe von Akten des Potsdamer Heeresarchivs durch die Sowjetunion. Militärgeschichtliche Mitteilungen 51 (1): 143–149.

Lorey, Wilhelm. 1916. Das Studium der Mathematik an den deutschen Universitäten seit Anfang des 19. Jahrhunderts. Abhandlungen über den mathematischen Unterricht in Deutschland. Vol. 3. Leipzig, Berlin: B. G. Teubner.

Lüdecke, Cornelia. 2002. Carl Ritters Lehrtätigkeit an der Allgemeinen Kriegsschule in Berlin 1820–1853. Berlin: Verlag für Wissenschafts- und Regionalgeschichte.

Lupfer, Timothy T. 1981. The Dynamics of Doctrine. Changes in German Tactical Doctrine During the First World War. Leavenworth Papers No. 4. Fort Leavenworth, Kan.: Combat Studies Institute, U.S. Army Command and General Staff College.

Macho, Thomas. 1996. Über Wittgenstein. In Wittgenstein. Ausgewählt und vorgestellt von Thomas Macho. Philosophie jetzt! 11–87. München: E. Diederichs.

Macrae, Norman. 1992. John von Neumann. New York: Pantheon Books.

Mahoney, Michael S. 1985. Diagrams and Dynamics. Mathematical Perspectives on Edgerton's Thesis. In Science and the Arts in the Renaissance, ed. John W. Shirley and F. David Hoeniger, 198–220. Washington, D.C.: Folger Shakespeare Library.

Maistrov, Leonid Efimovich. 1974. Probability Theory. A Histroical Sketch, ed. and trans. Samuel Kotz. New York and London: Academic Press.

Mancosu, Paolo. 1999. Between Vienna and Berlin. The immediate Reception of Gödel's Incompleteness Theorems. History and Philosophy of Logic 20:33–45.

Mancosu, Paolo. 2002. Phenomenology and Mathematics. Weyl at a crossroads. In Die Philosophie und die Wissenschaften. Zum Werk Oskar Beckers, ed. Jürgen Mittelstrass and Annemarie Gethmann-Siefert, 129–148. München: Wilhelm Fink.

Manegold, Karl-Heinz. 1966. Eine École Polytechnique in Berlin. In Technikgeschichte 33 (2): 182–196.

von Manstein, Erich. 1958. Aus einem Soldatenleben. 1887–1939. Bonn: Athenäum Verlag.

Manzano, María. 1997. Alonzo Church: His Life, His Work and Some of His Miracles. History and Philosophy of Logic 18:211–232.

Marose, Monika. 2005. *Unter der Tarnkappe. Felix Hartlaub. Eine Biographie.* Berlin: Transit.

McCarty, David. 2004. David Hilbert and Paul Du Bois-Reymond. Limits and Ideals. In One hundred years of Russell's paradox. Mathematics, logic, philosophy, ed. Godehard Link, 517–532. Berlin: Walter de Gruyter.

McCulloch, Warren. 1974. Recollections of the Many Sources of Cybernetics. American Society of Cybernetics Forum 6 (2): 5–16.

McGuinness, Brian. 1979. Editor's Preface. In Ludwig Wittgenstein and the Vienna Circle, Conversations Recorded by Friedrich Waismann, ed. Brian F. McGuinness, trans. Joachim Schulte and Brian F. McGuinness, 11–31. Oxford: Basil Blackwell.

McGuinness, Brian. 1988. Wittgenstein: A Life. Young Ludwig 1889–1921. London: Duckworth.

Mehrtens, Herbert. 1990. Moderne—Sprache—Mathematik. Eine Geschichte der Disziplin und des Subjekts formaler Systeme. Frankfurt am Main: Suhrkamp.

Mehrtens, Herbert. 1996. Mathematics and War. Germany 1900–1945. In National Military Establishments and the Advancement of Science and Technology. Studies in 20th-Century History, ed. Paul Forman and José Manuel Sanchez-Ron, 87–134. Boston Studies in Philosophy of Science, vol. 180. Dordrecht: Kluwer Academic Publishers.

Meier-Welcker, Hans. 1960. Unterricht und Studium in der Kriegsgeschichte angesichts der radikalen Wandlung im Kriegswesen. Wehrkunde 9: 608–612.

Messerschmidt, Manfred, ed. 1982. Die Militärgeschichte. Probleme-Thesen-Wege. Beiträge zur Militär- und Kriegsgeschichte, vol. 25. Stuttgart: Deutsche Verlags-Anstalt.

Miksche, Ferdinand Otto. 1976. Vom Kriegsbild. Stuttgart-Degerloch: Seewald.

Mirowski, Philip. 2002. Machine Dreams. Economics Becomes a Cyborg Science. Cambridge: Cambridge University Press.

Model, Hansgeorg. 1968. Der deutsche Generalstabsoffizier. Seine Auswahl und Ausbildung in Reichswehr, Wehrmacht und Bundeswehr. Frankfurt am Main: Bernard & Graefe.

Moll, Konrad. 1982. Von Erhard Weigel zu Christian Huygens. Studia Leibnitiana 14 (1): 56–72.

von Moltke, Helmut. 1911. Briefe über Zustände und Begebenheiten in der Türkei aus den Jahren 1835 bis 1839. Berlin: Mittler.

Monk, Ray. 1990. Ludwig Wittgenstein. The Duty of Genius. London: Jonathan Cape.

de Mora Charles, Maria Sol. 1992. Quelques jeux de hazard selon Leibniz (Manuscrits inédits). *Historia Mathematica* 19 (2): 125–158.

von Müffling, Karl. 1824. Anzeige. Militair-Wochenblatt 402: 2973.

Müller, Aloys. 1923. Über Zahlen als Zeichen. Mathematische Annalen 90:153–158.

Müller, Paul (alias Waldemar Gurian). 1943. Entscheidung und Ordnung. Zu den Schriften von Carl Schmitt. Schweizerische Rundschau 34:566–576.

von Neumann, John. 1927 (May 12). Letter to Trowbridge. International Education Board (IEB), series 1, subseries 3, box 55, folder 896, John L. Newmann/1926–1938. Rockefeller Archive Center.

von Neumann, John. 1928. Zur Theorie der Gesellschaftsspiele. Mathematische Annalen 100: 295–320. For an English transtation, see On the Theory of Games of Strategy. In Contributions to the Theory of Games, Vol. 4, ed. A.W. Tucker and R.D. Luce, trans. Sonya Bargmann, 13–42. Princeton: Princetion University Press, 1959.

von Neumann, John. 1983. The Formalist Foundations of Mathematics. In Philosophy of Mathematics: Selected Readings, ed. Paul Benacerraf and Hilary Putnam, trans. Erna Putnam and Gerald J. Massey. Cambridge: Cambridge University Press. Originally published as Die formalistische Grundlegung der Mathematik. In *Erkenntnis, Bericht über die 2. Tagung für Erkenntnislehre der exakten Wissenschaften in Königsberg 1930*, 2:2/3 (1931): 116–121.

Nicolin, Günther, ed. 1971. Hegel in Berichten seiner Zeitgenossen. Berlin: Akademie-Verlag.

Ore, Oystein. 1956. The Gambling Scholar. Princeton: Princeton University Press.

Ott, Eugen. 1965. Aus der Vorgeschichte der Machtergreifung des Nationalsozialismus' vor dem Rhein-Ruhr-Klub e.V. am 19 Mai 1965 in Düsseldorf, Nachlaß Carl Schmitt, RW 265–21410, Blatt 9. Landesarchiv Nordrhein-Westfalen, Hauptstaatsarchiv Düsseldorf.

Otto, Helmut. 1993. Das ehemalige Reichsarchiv. Streiflichter seiner Geschichte und der wissenschaftlichen Aufarbeitung des Ersten Weltkrieges. In Potsdam: Staat, Armee, Residenz in der preussisch-deutschen Militärgeschichte, ed. Bernhard R. Kroener, assisted by Heiger Ostertag, 421–434. Frankfurt am Main, Berlin: Propylaen.

Pachukanis, Eugen. 1970. Allgemeine Rechtslehre und Marxismus. Versuch einer Kritik der juristischen Grundbegriffe. Archiv sozialistischer Literatur. Vol. 3. Trans. Edith Hajós. Frankfurt am Main: Verl. Neue Kritik.

Page, Helena P. 1989. *General Friedrich Olbricht. Ein Mann des 20. Juli.* Bonn: Bouvier.

von Papen, Franz. 1952. *Der Wahrheit eine Gasse.* Munich: Paul List Verlag.

Paret, Peter. 1982. Kleist und Clausewitz: A Comparative Sketch. In *Festschrift für Eberhard Kessel zum 75. Geburtstag,* ed. Heinz Duchhardt and Manfred Schlenke, 130–139. Munich: Fink.

Paret, Peter. 1985. Clausewitz and the State: The Man, His Theories, and His Times. Princeton: Princeton University Press.

Petzold, Hartmut. 1992. Moderne Rechenkünstler—Die Industrialisierung der Rechentechnik in Deutschland. Munich: C. H. Beck.

Pichler, Alois. 2004. *Wittgensteins philosophische Untersuchungen. Vom Buch zum Album. Studien zur österreichischen Philosophie.* Vol. 36. Amsterdam: Rodopi.

Pieper, Herbert. 2003. Netzwerk des Wissens und Diplomatie des Wohltuns. Alexander von Humboldt, Carl Friedrich Gauß und Gustav Dirichlet, Jacob Jacobi, Eduard Kummer, Gotthold Eisenstein, Berliner Manuskripte zur Alexander-von-Humboldt-Forschung. Vol. 20. Berlin: Alexander-von-Humboldt-Forschungsstelle.

Post, Gaines. 1973. *The Civil-Military Fabric of Weimar Foreign Policy.* Princeton: Princeton University Press.

Poten, Bernhard. 1889. Reiswitz. Vol. 28. Allgemeine Deutsche Biographie, ed. historische Commission bei der königl. Akademie der Wissenschaften. Leipzig: Duncker & Humblot GmbH.

du Praissac, Sieur. 1639. The Art of Warre or Militarie Discourses. Cambridge: Printed by Roger Daniel, printer to that famous Uniuersitie.

Praun, Albrecht. 1951. Nachrichtenverbindungen bei Kriegsspielen und Rahmenuebungen. Attachment 3 of Rudolf Hofman, Über "Kriegsspiele." Bundesarchiv-Militärachiv, Bestand: P-094.

von Priesdorff, Kurt. 1937. Karl Friedrich von dem Knesebeck. Vol. 7. Soldatisches Führertum, ed. Kurt von Priesdorff, 344–348. Hamburg: Hanseatische Verlagsanstalt.

Pyta, Wolfram. 1992. Vorbereitungen für den militärischen Ausnahmezustand unter den Regierungen Papen / Schleicher. Militärgeschichtliche Mitteilungen 51 (2): 385–428.

Pyta, Wolfram. 1998. Verfassungsumbau, Staatsnotstand und Querfront: Schleichers Versuche zur Fernhaltung Hitlers von der Reichskanzlerschaft August 1932 – Januar 1933. In Gestaltungskraft des Politischen. Festschrift für Eberhard Kolb, ed. Wolfram Pyta and Ludwig Richter, 173–197. Berlin: Duncker & Humblot.

Pyta, Wolfram. 1999. Konstitutionelle Demokratie statt monarchischer Restauration. Die verfassungspolitische Konzeption Schleichers in der Weimarer Staatskrise. Vierteljahrshefte fur Zeitgeschichte 47:417–441.

Pyta, Wolfram and Gabriel Seiberth. 1999. Die Staatskrise der Weimarer Republik im Spiegel des Tagebuchs von Carl Schmitt. Der Staat 38:423–448, 594–610.

Radbruch, Kurt. 1997. Mathematische Spuren in der Literatur. Darmstadt: Wissenschaftl. Buchges.

von Reiche, Ludwig. 1857. Memoiren des königlich preußischen Generals der Infanterie, Erster Theil: Von 1775 bis 1814, ed. Louis von Weltzien. Leipzig: Otto Wigand.

Reichenbach, Hans. 1929. Stetige Wahrscheinlichkeitsfolgen. Zeitschrift fur Physik 53:274–307.

von Reiswitz, Georg Heinrich Rudolf Johann. 1824. Anleitung zur Darstellung militairischer Manöver mit dem Apparat des Kriegs-Spiels. Berlin.

von Reiswitz, George Leopold. 1812. Taktisches Kriegs-Spiel oder Anleitung zu einer mechanischen Vorrichtung um taktische Manoeuvres sinnlich darzustellen. Berlin.

von Reiswitz, George Leopold. 1816. Literarisch-kritische Nachrichten über die Kriegsspiele der Alten und Neuern. Marienwerder.

Ricciardi, Franc M., Clifford J. Craft, Donald G. Malcolm, Richard Bellman, Charles Clark, Joel M. Kibbee, and Richard H. Rawdon. 1957. In Top Management Decision Simulation: The AMA Approach, ed. Elizabeth Marting. New York.

Richardson, R. G. D. 1942. Letter to Douglas C. Jackson, November 19, 1942. The Institute Archives and Special Collections, NWP, box 4, folder 62, MIT Libraries.

Ries, Adam. 1522. Rechenung auff der linihen vnd federn. Erfurt: Mathes Maler.

Riley, Vera, and John P. Young. 1957. Bibliography on War Gaming. Chevy Chase, Md.: Operations Research Office, The Johns Hopkins University.

Robb, Graham. 1994. Balzac. A Biography. London: Macmillan Publishers Limited.

Röller, Nils. 2000. Medientheorie im epistemischen Übergang. Hermman Weyls Philosophie der Mathematik und Naturwissenschaften und Ernst Cassirers Philosophie der symbolischen Formen im Wechselverhältnis. Weimar: VDG.

Rotman, Brian. 1987. Signifying Nothing. The Semiotics of Zero. New York: St. Martin's Press.

Schäffner, Wolfgang. 1991. Psychiater machen mobil durch Arbeit und Kriegsspiel. Zur Allianz von Militär und Psychiatrie. Messungen. Zeitschrift für Interpretationswissenschaften 1:25–33.

von Scharfenort, Louis A. 1910. Die Königlich Preussische Kriegsakademie. 1810–1910. Berlin: Mittler.

Scharlau, Winfried, and Eberhard Knobloch. 1989. Berlin. Universität. In Mathematische Institute in Deutschland 1800–1945. Dokumente zur Geschichte der Mathematik. Vol. 5, 25–48. Braunschweig, Wiesbaden: Friedr. Vieweg & Sohn.

von Scharnhorst, Gerhard Johann David. 1973. *Nutzen der militärischen Geschichte; Ursach ihres Mangels. Ein Fragment aus dem Scharnhorst-Nachlass*, Faksimilie d. Handschrift mit Übertragung und Einführung v. Ursula von Gersdorff. Osnabrück: Biblio-Verlag.

von Schellendorff, Paul Bronsart. 1875. Der Dienst des Generalstabes. Vol. 1. Berlin: E. S. Mittler.

Schmitt, Carl. 1911. Der Adressat. Die Rheinlande 11 (12): 429–430.

Schmitt, Carl. 1912. Richard Wagner und eine neue "Lehre vom Wahn." *Bayreuther Blätter* 35:239–241.

Schmitt, Carl. 1913. Juristische Fiktionen. Deutsche Juristenzeitung 18 (12): 804–805.

Schmitt, Carl. 1934. Der Führer schützt das Recht. Deutsche Juristen-Zeitschrift 39 (15): columns 945–950.

Schmitt, Carl. 1936. Politik. Vol. 1. Handbuch der neuzeitlichen Wehrwissenschaften: Wehrpolitik und Kriegsführung, ed. on behalf of Deutschen Gesellschaft für Wehrpolitik und Wehrwissenschaften und unter Mitarbeit umstehend aufgeführter Sachverständiger von Hermann Franke, 549. Berlin, Leipzig: Walter de Gruyter.

Schmitt, Carl. 1952. Foreword. In Hamlet. Sohn der Maria Stuart by Lilian Winstanley, 7–25. Trans. Anima Schmitt. Pfullingen: Verlag Günther Neske.

Schmitt, Carl. 1991. *Völkerrechtliche Großraumordnung mit Interventionsverbot für raumfremde Mächte. Ein Beitrag zum Rechtsbegriff im Völkerrecht*. Berlin: Duncker & Humblot.

Schmitt, Carl. 1999. Letter to Ernst Jünger, September 17, 1941. In Ernst Jünger—Carl Schmitt. Briefe 1930–1983, ed. Helmuth Kiesel, 128–130. Stuttgart: Klett-Cotta.

Schmitt, Carl. 2005. Political Theology: Four Chapters on the Concepts Of Sovereignty, trans. George Schwab. Chicago: University of Chicago Press. Originally published as *Politische Theologie*. München: Duncker & Humblot, 1922.

Schmitt, Carl. 1956. Hamlet oder Hekuba. Der Einbruch der Zeit in das Spiel. Düsseldorf: Klett-Cotta. For an English translation, see Hamlet or Hecuba: The Intrusion of the Time into the Play, trans. David Pan and Jennifer R. Rust. New York: Telos Press, 2009.

Schneider, Ivo. 1993. Johannes Faulhaber 1580 –1635. Rechenmeister in einer Welt des Umbruchs. Basel: Birkhäuser.

Schnelle, Helmut. 1962. Zeichensysteme zur wissenschaftlichen Darstellung. Ein Beitrag zur Entfaltung der Ars characteristica im Sinne von G. W. Leibniz. Stuttgart-Bad Cannstatt: Frommann-Holzboog.

Scholz, Heinrich. 1931. Geschichte der Logik, Geschichte der Philosophie in Längsschnitten 4. Berlin: Junker u. Dünnhaupt.

Scholz, Heinrich. 1920. Zum "Untergang des Abendlandes." Eine Auseinandersetzung mit Oswald Spengler. Berlin: Reuther & Reichard.

Schottelius, Justus Georg. 1991. Der schreckliche Sprachkrieg. Horrendum Bellum Grammaticale, ed. Friedrich Kittler and Stefan Rieger. Leipzig: Reclam.

Selenus, Gustavus (Alias Herzog August II) von Braunschweig-Lüneburg. 1978. Das Schach- oder Königsspiel. Reprint from 1616, ed. Viktor Kortschnoi and Klaus Lindörfer. Zürich: Edition Olms.

Sembdner, Helmut, ed. 1957a. Heinrich von Kleists Lebensspuren. Dokumente und Berichte der Zeitgenossen. Bremen: Carl Schünemann Verlag.

Sembdner, Helmut. 1957b. Heinrich und Marie von Kleist. Jahrbuch der deutschen Schillergesellschaft 1:157–178.

Shannon, Claude E. 1949. Communication Theory of Secrecy Systems. Bell System Technical Journal 28:656–715.

Shannon, Claude E. 1950a. A Chess-Playing Machine. Scientific American 182 (2): 2124–2133.

Shannon, Claude E. 1950b. Programming a Computer for Playing Chess. Philosophical Magazine, series 7, 41 (314): 256–275.

Shannon, Claude E. 1955. Game Playing Machines. Journal of the Franklin Institute 260 (6): 447–453.

Siegert, Bernhard. 1996. L'Ombra della macchina alata. Gabriele d'Annunzios ‚renovatio imperii' im Licht der Luftkriegsgeschichte 1909–1940. In Der Dichter als Kommandant. D'Annunzio erobert Fiume, ed. Hans Ulrich Gumbrecht, Friedrich Kittler, and Bernhard Siegert, 261–305. München: Wilhelm Fink.

Siegert, Bernhard. 2003. Passage des Digitalen. Zeichenpraktiken der neuzeitlichen Wissenschaften 1500–1900. Berlin: Brinkmann & Bose.

Smith, Adam. 2003. History of Astronomy. In Essays on Philosophical Subjects, ed. William P. D. Wightman and J. C. Bryce, 33–105. Oxford: Oxford University Press.

Speer, Albrecht. 1996. Erinnerungen. Frankfurt am Main, Berlin: Ullstein.

Spengler, Oswald. 1926. The Decline of the West: Form and Actuality. trans. Charles Francis Atkinson. Vol. 1. New York: Oxford University Press.

Spieß, Alfred, and Heiner Lichtenstein. 1979. *Das Unternehmen Tannenberg*. Munich: Limes Verlag.

Stahl, Friedrich-Christian. 1977. Die Organisation des Heeresarchivwesens in Deutschland 1936–1945. Schriften des Bundesarchivs 25:69–101.

van Stigt, Walter P. 1990. Brouwer's Intuitionism. Studies in the History & Philosophy of Mathematics. Amsterdam: North-Holland.

Stock, Eberhard. 1987. Die Konzeption einer Metaphysik im Denken von Heinrich Scholz. Theologische Bibliothek Töpelmann 44. Berlin: de Gruyter.

Stone, Norman. 1975. The Eastern Front 1914–1917. New York: Scribner.

Struve, Horst, and Rolf Struve. 1997. Leibniz als Wahrscheinlichkeitstheoretiker. *Studia Leibnitiana* 29 (1): 112–122.

Syed, Renate. 2001. Kanauj, Maukharis und das Caturanga. Der Ursprung des Schachspiels und sein Weg von Indien nach Persien. Kelkheim: Foerderkreis Schach-Geschichtsforschung.

Taubes, Jacob. 1987. *Ad Carl Schmitt. Gegenstrebige Fügung*. Berlin: Merve Verlag.

von Taysen, Adalbert. 1936. Führerausbildung. Vol. 1. Handbuch der neuzeitlichen Wehrwissenschaften: Wehrpolitik und Kriegsführung, ed. im Auftrage der Deutschen Gesellschaft für Wehrpolitik und Wehrwissenschaften und unter Mitarbeit umstehend aufgeführter Sachverständiger von Hermann Franke, 97–102. Berlin: Walter de Gruyter.

Turing, Alan. 1937. On computable numbers, with an application to the Entscheidungsproblem. Proceedings of the London Mathematical Society, series 2, vol. 42 (1936–1937): 230–265. Corrections published in Proceedings of the London Mathematical Society, series 2, vol. 43: 544–546.

Turing, Alan. 1953. Chess. In Faster Than Thought: A Symposium on Digital Computing Machines, ed. Bertram Vivian Bowden, 286–295. London: Pitman.

Ueberschär, Gerd. 1998. Die militärische Planung für den Angriff auf die Sowjetunion. In Der deutsche Angriff auf die Sowjetunion 1941, ed. Gerd Ueberschär and Lev A. Bezymenskij, 21–37. Darmstadt: Primus.

Ulam, Stanislaw. John von Neumann, 1903–1957. Bulletin of the American Mathematical Society 64 (8), pt. 2: 1–49.

Upton-Ward, Judith M. 1992. The Rule of the Templars: The French Text of the Rule of the Order of the Knights Templar/Templars. Woodbridge: Boydell Press.

van Creveld, Martin. 1982. Fighting Power: German and U.S. Army Performance, 1939–1945. Westport, Conn.: Greenwood Press.

van Dalen, Dirk. 1999. Mystic, Geometer, and Intuitionist. The Life of L.E.J. Brouwer. Oxford: Oxford University Press.

von Verdy du Vernois, Julius. 1881. *Beitrag zum Kriegsspiel*. Berlin: E.S. Mittler.

Villinger, Ingeborg. 1992. Politische Fiktionen. Carl Schmitts literarische Experimente. In *Technopathologien*, ed. Bernhard Dotzler, 191–222. Munich: Fink.

Virilio, Paul. 2005. Negative Horizon: An Essay in Dromoscopy, trans. Michael Degener. London: Continuum. Originally published as *L'horizon négatif: essai de dromoscopie*. Paris: Editions Galilée, 1984.

Vogelsang, Thilo. 1962. *Reichswehr, Staat und NSDAP. Beiträge zur deutschen Geschichte 1930–1932*. Document 38. Stuttgart.

Voigt, Gerhard. 1975. Goebbels als Markentechniker. In Warenästhetik. Beiträge zur Diskussion, Weiterentwicklung und Vermittlung ihrer Kritik, ed. Wolfgang Fritz Haug, 231–260. Frankfurt am Main: Suhrkamp.

Voisé, Wlademar. 1967. Leibniz' Model of Political Thinking. Organon 4: 187–195.

Voltmer, Ernst. 1988. Standart, Carroccio, Fahnenwagen. Zur Funktion der Feld- und Herrschaftszeichen mittelalterlicher Städte am Beispiel der Schlacht von Worringen 1288. Blätter für deutsche Landesgeschichte 127:187–209.

Vossen, Peter. 1962. Der Libellus Scolasticus des Walther von Speyer. Ein Schulbericht aus dem Jahre 984. Berlin: Walter von Speyer.

Waismann, Friedrich. 1931. Vorbemerkung. Erkenntnis. Bericht über die 2, Tagung für Erkenntnislehre der exakten Wissenschaften in Königsberg 1930 2 (2–3): 87.

Waismann, Friedrich. 1979. Wittgenstein and the Vienna Circle: Conversations recorded by Friedrich Waismann, ed. Brian McGuinness, trans. Joachim Schulte and Brian McGuinness. Oxford: Blackwell. Originally published as *Ludwig Wittgenstein and the Vienna Circle, Conversations Recorded by Friedrich Waismann*, ed. Brian F. McGuinness, trans. Joachim Schulte and Brian F. McGuinness. Oxford: Basil Blackwell, 1979.

Weickmann, Christoph. 1664. New-erfundenes grosses Königs-Spiel etc. Ulm: Kühne.

Weigel, Erhard. 2004. Arithmetische Beschreibung der Moral-Weißheit von Personen und Sachen. Vol. 2. Werke. Clavis pansophiae 3, ed. Thomas Behme. Stuttgart-Bad Cannstatt: Frommann-Holzboog.

Weyl, Hermann. 1924. Randbemerkungen zu Hauptproblemen der Mathematik. *Mathematische Zeitschrift* 20:131–150.

Weyl, Hermann. 1925. Die heutige Erkenntnislage in der Mathematik. Symposium 1 (1): 1–23.

Weyl, Hermann. 2009. Philosophy of Mathematics and Natural Science. Princeton: Princeton University Press. Originally published as *Philosophie der Mathematik und Naturwissenschaft*. Munich: Oldenbourg, 1927.

Weyl, Hermann. 1968a. Über den Symbolismus der Mathematik und mathematischen Physik. In Gesammelte Abhandlungen. Vol. 4. Ed. Komaravolu Chandrasekharan, 527–536. Berlin: Springer.

Weyl, Hermann. 1968b. Über die neue Grundlagenkrise der Mathematik. Vol. 2. Gesammelte Abhandlung, ed. Komaravolu Chandrasekharan, 143–180. Berlin: Springer-Verlag.

Winstanley, Lilian. 1952. Hamlet. Sohn der Maria Stuart. Pfullingen: Verlag Günther Neske. Originally published as *Hamlet and the Scottish Succession. Being an Examination of the Relations of the Play of Hamlet to the Scottish Succession and the Essex Conspiracy*. Cambridge: University Press, 1921.

Winterbotham, Frederick W. 1974. The Ultra Secret. London: Weidenfeld and Nicolson.

Wittgenstein, Ludwig. 1961. Notebooks 1914–1916, ed. Georg Henrik von Wright and Gertrude Elizabeth Margaret Anscombe, trans. G. E. M. Anscombe. Oxford: Blackwell.

Wittgenstein, Ludwig. 1974. Philosophical Grammar, ed. Rush Rhees, trans. Anthony Kenny. Berkeley: University of California Press. For the original text, see German *Philosophische Grammatik*. Vol. 4, *Werkausgabe*, ed. Rush Rhees. Frankfurt am Main: Suhrkamp, 1984.

Wittgenstein, Ludwig. 1979a. Calculus and Application. In Ludwig Wittgenstein and the Vienna Circle, Conversations Recorded by Friedrich Waismann, ed. Brian F. McGuinness, trans. Joachim Schulte and Brian F. McGuinness, 170. Oxford: Basil Blackwell, 1979a. For the original text, see Kalkül und Anwendung. Vol. 3, *Werkausgabe: Ludwig Wittgenstein und der Wiener Kreis. Gespräche, aufgezeichnet von Friedrich Waismann*, ed. Brian. F. McGuinness, 170. Frankfurt am Main: Suhrkamp, 1989.

Wittgenstein, Ludwig. 1979b. Consistency III. In Ludwig Wittgenstein and the Vienna Circle, Conversations Recorded by Friedrich Waismann, ed. Brian F. McGuinness, trans. Joachim Schulte and Brian F. McGuinness, 119–121. Oxford: Basil Blackwell. For the original text, see Widerspruchsfreiheit III. Vol. 3, *Werkausgabe: Ludwig Wittgenstein und der Wiener Kreis. Gespräche, aufgezeichnet von Friedrich Waismann*, ed. Brian F. McGuinness, 119–121. Frankfurt am Main: Suhrkamp, 1989.

Wittgenstein, Ludwig. 1979c. What to Say at Konigsberg. In Ludwig Wittgenstein and the Vienna Circle, Conversations Recorded by Friedrich Waismann, ed. Brian F. McGuinness, trans. Joachim Schulte and Brian F. McGuinness, 102–107. Oxford: Basil Blackwell. For the original text, see Was in Königsberg zu sagen wäre. Vol. 3, *Werkausgabe: Ludwig Wittgenstein und der Wiener Kreis. Gespräche, aufgezeichnet von*

Friedrich Waismann, ed. Brian F. McGuinness, 102–107. Frankfurt am Main: Suhrkamp, 1989.

Wittgenstein, Ludwig. 1980. Wittgenstein's Lectures—Cambridge. 1930–1932. From the Notes of John King and Desmond Lee, ed. Desmond Lee. Oxford: Basil Blackwell, 1980.

Wittgenstein, Ludwig. 1998. Culture and Value: A Selection from Posthumous Remains, ed. Georg H. von Wright in collaboration with Heikki Nyman, trans. Peter Winch. Oxford: Blackwell Publishing. For the original text, see Vermischte Bemerkungen. Vol. 8, *Werkausgabe*. Frankfurt am Main: Suhrkamp, 1989.

Wittgenstein, Ludwig. 1991. In *Geheime Tagebücher 1914–1916*, ed. Wilhelm Baum. Wien: Turia & Kant.

Wittgenstein, Ludwig. 2001. Tractatus Logico-Philosophicus, trans. David F. Pears and Brian F. McGuinness. London: Routledge. Originally published as Logisch-philosophische Abhandlung. Tractatus logico-philosophicus. *Annalen der Naturphilosophie* 14 (1921). Also in vol. 1, *Werkausgabe*, 8–85. Frankfurt am Main: Suhrkamp, 1995.

Wittgenstein, Ludwig. 2009. Philosophical Investigations, trans. G. E. M. Anscombe, P. M. S. Hacker, and Joachim Schulte. Oxford: Blackwell Publishing. Originally published as Philosophische Untersuchungen. Vol. 1, Wittgenstein, *Werkausgabe*, 225–580. Frankfurt am Main: Suhrkamp, 1989.

von Wright, Georg Henrik. 1955. Ludwig Wittgenstein: A Biographical Sketch. Philosophical Review 4 (64): 527–545.

Zabecki, David T. 1994. Steel Wind. Colonel Georg Bruchmüller and the Birth of Modern Artillery. The Military Profession. Westport, Conn.: Praeger.

Zermelo, Ernst. 1913. Über eine Anwendung der Mengenlehre auf die Theorie des Schachspiels. Proceedings of the Fifth International Congress of Mathematicians, Cambridge, 1912, ed. E. W. Hobson and A. E. H. Love, vol. 2, 501–504.

Zuse, Konrad. 1972. Kommentar zum Plankalkül. In Der Plankalkül, ed. Gesellschaft für Mathematik und Datenverarbeitung, no. 63, 1–35. St. Augustin: Ges. für Mathematik und Datenverarbeitung.

Zuse, Konrad. 1986. Der Computer, mein Lebenswerk. Berlin: Springer.

Index

Printed in the United States
by Baker & Taylor Publisher Services